Construction and the Natural Environment

Construction and the Natural Environment

R. M. Lowton

BUTTERWORTH
HEINEMANN

Butterworth-Heinemann
Linacre House, Jordan Hill, Oxford OX2 8DP
A division of Reed Educational and Professional Publishing Ltd

 A member of the Reed Elsevier plc group

OXFORD   BOSTON   JOHANNESBURG
MELBOURNE   NEW DELHI   SINGAPORE

First published 1997

© R. M. Lowton 1997

British Library Cataloguing in Publication Data
Lowton, R. M.
   Construction and the natural environment
   1. Construction industry – Environmental aspects   2. Building
   – Environmental aspects
   I. Title
   690

ISBN 0 7506 2302 0

Composition by Genesis Typesetting, Laser Quay, Rochester, Kent
Printed and bound in Great Britain by Scotprint Ltd, Musselburgh

Contents

Introduction

The environment created by the construction industry is designed to improve the quality of our time on this planet. We spend by far the greater proportion of our lives inside the built environment.

It is evident that in the past the construction industry has regarded the natural environment with disregard, using mineral resources, land and energy with consideration only for profit. Recent national and EU regulations are aimed at a reduction in environmental damage. There is a greater public demand for environmentally friendly products now than there has ever been before.

The responsibility of pollution through energy use is laid at the door of the user, but the user has no alternative but to live or work in the building as it was designed. Insufficient consideration of the internal environment leads to buildings which are expensive to heat and which provide an unpleasant home or workplace. Insufficient consideration of the natural environment, which is the source of our materials, results in a waste of resources, pollution problems and the associated costs.

This study does not propose that all construction technicians and professionals join environmental pressure groups. We are being made more aware by events, of the almost religious belief that if we work with nature everything will tick along nicely. As our demands from the natural environment multiply it is temporarily damaged. However, the natural environment works like a living organism. If we scar the landscape with quarries and open cast mines, in time plants will grow followed by insects, rabbits, foxes and so on. A polluted atmosphere is rejuvenated by natural diffusion. However, the 4000 or so Londoners who died as a result of atmospheric pollution in 1953 cannot be rejuvenated. We damage the environment at our peril.

The absence of a proper understanding of the natural environment has cost construction companies and the tax-payer millions of pounds each year. Many of the larger companies now invest in environmental policies.

This book shows that the financial and social costs involved in protecting the environmental system are not that great when compared to the cost of the alternative.

The scope of this book is GNVQ Construction and the Built Environment at Intermediate and Advanced level.

Many of the subject changes from First and National Diploma to GNVQ Intermediate and Advanced are associated with the increased requirement for technicians and professionals in the construction industry to consider the natural environment.

My primary aim is to provide lecturers and students with material which deals with the new environmental impetus in society generally, and the construction industry specifically.

The majority of the topics are taken from the GNVQ Intermediate and Advanced syllabuses. Additionally topics which were traditionally covered at National Diploma level and are absent from the GNVQ Intermediate and Advanced have been included for the sake of completeness.

1

Urbanization, public health and planning

The aims of this chapter are to:

1 Introduce the principle that early buildings were constructed solely for protection from elements of the natural environment.
2 Show that early urbanization was controlled by environmental factors.
3 Show that planning and public health are tied to considerations of environmental protection.

1.1 INTRODUCTION – A BRIEF HISTORY

The evidence for human activity in Britain before the last glacial stage is generally limited to artifacts, such as hand axes and pottery. For much of the duration of the glacial stages Britain was uninhabitable. As the climate became warmer following the last glacial stage the human population began to grow.

Ten thousand years ago Britain was covered by forest. Some was cleared by ancient peoples for agricultural purposes. As the bronze and iron age technologies advanced communities increased in size and more non-farming jobs were created.

The first British evidence of a planned village layout dates back to the Iron Age. However, evidence for town planning in a modern sense, comes from the Roman era. The roads within the towns were set out in a grid system with properly constructed storm water drains. Churches and entertainment centres were at the heart of the towns and graveyards outside the town boundary.

Following the termination of Roman dominance, the Saxons gradually overpowered the Romano-British and Celtic peoples over the next 200 years. The Anglo Saxons sectioned the land into estates and parishes so that it could be farmed efficiently. When the Normans invaded, overall control was passed to a new aristocratic class who paid rent to the monarch.

As the towns expanded, especially around the coast, buildings were constructed with little consideration for planning or hygiene. There was no real attempt to dispose of waste. It was left to the individual, who usually threw it into the street creating the greatest health hazard this country had yet seen. It is clear from the following case study that local authority control of planning must include provision for the disposal of waste.

Case study – bubonic plague

Our fear of rats probably stems from the Black Death (Bubonic plague), which came to Britain via the ports of Dorset and Hampshire in 1348. The black rats thrived on waste. They carried fleas which in turn carried bacteria in their guts. When the rats died the fleas transferred their attention, and the bacteria, to any available host. The majority of humans bitten by the flea died in agony within a few days.

Black Death is often considered a problem of the fourteenth century. However there have been several recurrences. In 1665, 20 per cent of Londoners died of it. Bubonic plague has appeared every few years somewhere in the world. The last two deaths in Britain were in Wales in 1900. Many archaeologists believe that Black Death was so severe in some villages that most perished and the remaining few were unable to maintain their industries and left.

In 1666 following the Great Fire of London, Parliament issued regulations for its rebuilding. However, like much legislation it was a reaction to a disaster rather than pre-emptive, positive planning. The government objective was to limit fires rather than control building. Furthermore, the legislation was implemented by architects rather than a government body, so control was still in the hands of the individual or business rather than the state.

A decline in the use of timber as a fuel through forest management and clearance, and the effects of the little ice age (approximately 1500–1900), increased the demand for coal. Small-scale coal mining had been around for centuries, but only produced enough for local domestic use. The scale of mining activities increased dramatically to fuel the energy intensive processes of the industrial revolution. Movement of people to the cities was inevitable. Manchester's population grew from 95 000 to 238 000 in twenty years from 1801. A rapid growth of the chemical industries followed to service the needs of textile and construction companies. Britain experienced unprecedented development, and yet there were still few state or local authority planning controls. As urbanization escalated more of the countryside was destroyed by mining, construction and pollution. The amount of land required to sustain urban communities is considerably greater than the area occupied by towns and cities. Industries require raw materials gained from mining, and people need food from agricultural land.

Britain is not a large country by any means, but it is more densely populated than many other countries of the world with around 230 people per km^2 compared with 11 people per km^2 in New Zealand and 1.8 per km^2 in Australia. It is probably the case that Britain is very close to a maximum sustainable population density. The population explosion since the 1800s is slowing so that it is now almost stable, and yet, of the twenty-four million hectares, only just over three million is urban development.

1.2 URBANIZATION

Urbanization is the changing of the character of an area from rural to urban. However because this

involves an increasing proportion of people living in towns and cities urbanization has come to be associated with relative population densities rather than the physical appearance of an area.

1.2.1 Determinism and possibilism

The relationship between man and the natural environment can be viewed as a matter of response. Geographers call it **determinism**. It implies that our behaviour is a response to the natural environment. The desire for shelter was partly in response to **climatic conditions**. The construction processes of early man were determined by local **geology** and **timber** for building materials. The **topography** determined where the shelter, or village was to be constructed. A change from hunting to farming resulted in the introduction of permanent settlements. The condition of the **soil** and presence of **surface water** determined whether the shelter was permanent or temporary. It also determined the type of plants and animals that could be kept and so established diet content.

The earliest artificial shelters were constructed using whatever materials were available locally and served simply to isolate their inhabitants from the weather. Building materials included local rock, branches, turfs, straw and mud. The earliest wattle and daub walls (Figure 1.1) are believed to have consisted of rough stakes driven into the ground and the gaps between filled with interlocking twigs, or wattle.

To keep out the weather the wattle was then daubed with mud or clay. Where larger timbers were available buildings were often constructed of jointed frames with a wattle and daub infill. If timber was more difficult to come by walls were constructed of stone giving the obvious advantages of longevity and

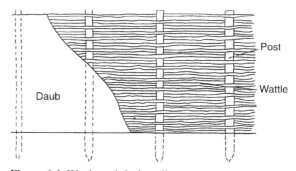

Figure 1.1 Wattle and daub wall

strength. The roof may have been covered with branches for structure and a layer of straw or turf or twigs for protection from rain. The practice of using materials from the local natural environment is still in evidence. Many Scottish towns are from local hard rock. London's houses are from clay bricks, and Bath is dominated by limestone buildings.

There is no question that the decisions and behaviour of early man were, in the main, developed in response to the local environmental conditions.

The opposite view is **possibilism** which suggests that choice always exists. For instance, in the design of buildings the architect is not constrained by local materials. In the past a house constructed of local limestone would often have oak lintels or stone arches if no suitably sized slabs of rock could be found locally. Now an architect can stipulate the heads of openings using whatever method or material is considered appropriate for that project (provided that the transport costs do not inflate the price of the component excessively). If the material is not local it will be manufactured or imported. The extent to which the construction of our built environment is the result of determinism or possibilism changes with national development and wealth. The population of a poor nation are forced by circumstances to use local materials for construction, to eat food which is produced locally and to live by social standards which are applicable to, and governed by, the local natural environment.

It seems then that wealth leads to possibilism while determinism results from poverty. However, changing the topography so that hillsides become terraces on which to grow food, the very act of construction, and the irrigation of land from deep wells may all be classified as the result of possibilism, as people have chosen to realize the potential of the world around them rather than be limited by existing environmental conditions.

The examples given above also show that when we use the natural environment we change it. However, the past has taught us some severe lessons. Environmentally aware industries make great efforts to minimize the change. Chapter 5 shows how water authorities attempt to return water to the natural environment in, as far as is possible, the same condition as when they removed it. This is not done simply to satisfy a 'Green lobby'. The water may be abstracted downstream and used again for domestic, recreational or industrial purposes. Destroying that facility may result in additional costs to the company, or heavy fines, or, since water is now managed by private companies, loss of licence. In effect we have no choice – we have to protect the natural environment.

1.2.2 Settlements

The location and construction of early settlements was determined by the available natural resources. The most important resource was, and still is, water. The extent of urban development has in many cases also been a function of the volumes of water available. Large rivers tend to have large settlements along them, not only because water is used for drinking, irrigation and cleaning, but also as a means of transport. An extension of this is the theory suggested that large areas of high ground are not favoured where lower ground is available. Modern substantiation from the London area supports this hypothesis. Figure 1.2 shows the River Thames and the River Kennet from central southern England.

There is a corridor of dense population which becomes more dense as the river becomes more navigable and is then reduced as the river water becomes saline and of less use to the basic requirements of early settlements.

A plentiful supply of timber would constitute a means of construction in the early settlements, and a source of warmth. The first buildings of early man were from timber.

The original use of rock, or stone, was a matter of finding stones of similar sizes and filling any gaps with mud. Dressing or cutting stone dates back at least 5000 years to ancient Egypt. One thousand years later almost any rock could be cut to almost any dimension. This art was copied and improved by the Greeks and the Romans. In hot dry climates it was discovered that moistened clay was easily moulded and dried hard in the sun. The first evidence of organized brick making dates back almost nine thousand years.

It comes as no surprise then, to find that Britain's largest city was originally a group of settlements in the valley of a large river, in an oak forest resting on a clay deposit.

1.2.3 Settlement classification

Permanent settlements are classified in many ways by different people. The terms **rural** and **urban** are not definitive. The former generally infers either a small town, village or hamlet and the associated industries. The latter larger towns and cities and their activities. Generally the simplest definition is by size. It is possible to account for size in the area of land covered by the settlement, but this makes no allowance for population density. The commonest method is therefore by population. Table 1.1 shows how settlements are classified by population.

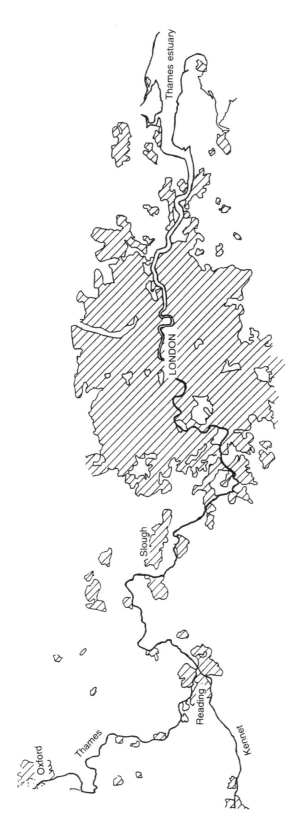

Figure 1.2 Part of the Thames Valley corridor

Table 1.1 Settlement type by population

Type of settlement	Maximum population
Hamlet	100
Small village	500
Large village	2 000
Small town	10 000
Large town	100 000
City	1 000 000
Conurbation	10 000 000

There are, of course, problems associated with this classification system. One is the classification of cities. Southampton became a city in 1964, and was previously one of the biggest towns on the south coast of England with a population of around 205 000, while the nearby city of Salisbury has a population of around 36 000 (1995). Another problem is local perception. The most important settlement in an area is considered locally to be more than a village. Fort William in Scotland may seem to the locals a large town, while Crawley in Sussex, falling between London and Brighton, is a small to medium sized town. In reality Crawley is many times larger in terms of both population and area than Fort William.

Settlements may also be classified by function, or the perceived function, which again shows the difficulty architects have in designing an appropriate and acceptable style. Any settlement with new town or city in its title serves the function of an **overspill development**. Oxford and Cambridge are often referred to as **university towns** while Oxford has long been a major engineering city to the motor industry. Blackpool is a **holiday town** and Clovelly in Devon, a **holiday village**. Sheffield is a **steel town**, Grimsby a **fishing town** and Nottingham a **mining town**. These perceptions in many ways determine what is expected of their development.

Where a settlement is classified by function the development of new buildings may clash with the evolution of the settlement in response to its environment. One such example is the Tricorn Centre in Portsmouth, historically an island city with a close relationship with the sea. The Tricorn Centre was constructed close to the dockyard and the city centre, and on a junction of two roads which lead to different parts of the old town (an area generally considered to be of character and charm). The construction of the Tricorn Centre, finished in ribbed undecorated concrete and plastic sheeting, in such an area was in stark contrast to development of the city through its

perceived environmental evolution. It has been insulted by Portsmouth people since its construction and was once voted the ugliest building in Europe. It may be more appropriate to describe it as the most out of place building in Europe.

There is no doubt that settlements may be classified, but there is often a degree of subjectivity.

1.2.4 Urban growth and communication

The effects of the increase in the size, density and spread of structures is often considered detrimental to both the natural environment and the expanding settlement. The expansion of a town or the construction of a motorway are often at the expense of agricultural land or areas of natural beauty. Upward growth, in the form of taller buildings, creates shadows and wind eddies and associated social problems. Whether the town expands outward or upward there are always traffic planning considerations. The increased use of existing road systems in built-up areas causes severe traffic problems, such as noise, air pollution and driver frustration, especially in the centre of the city. This is often associated with a depopulation of the town and city centres, where the larger properties are converted to offices and the smaller are demolished and redeveloped.

The other option is to develop towns and cities outward. This demands the construction of many miles of roads over a larger area which costs considerably more in the short term and increases maintenance costs in the future.

1.2.5 Slums

The lack of building control was not a real issue until the onset of the industrial revolution. There were two important factors which influenced population densities. The first was the constant discoveries of ores, minerals and rocks which gave Britain the lead in so many areas of eighteenth and nineteenth century technology. The population drifted towards the industrial areas because that was where the money was. The comparatively huge volume of exports caused population movement to dock cities such as London Liverpool and Bristol. The backbone of Britain's workforce prior to the mid-eighteenth century was predominantly agricultural. By the early-nineteenth century many of the agricultural workers had left the country to live in the cities. (A mass migration from the countryside to the cities is, however, a misconception.)

The second factor to influence population densities in cities was a result of an improvement in health care at the beginning of the twentieth century. This resulted in a reduction in child mortality and an increase in the avearge life span.

The working classes moved closer to the docks, factories and mines, while the upper and middle classes were able to choose where they lived and choose a setting which more closely resembled the natural environment. The poor could not afford to rent decent accommodation. Most of the rooms in large houses were let to families at about 3 pence per night. This obviously caused sanitation problems. Tapwater was often supplied by a standpipe in the road serving hundreds of people. One reporter found that, of seven people living in a room, one had diphtheria or consumption and the latest offspring was lying dead in the corner. This was considered commonplace. Over half of the children died before they reached five years, and those who survived were often mentally and physically underdeveloped.

With a greatly segregated society based on capital and wealth the rich got richer and the poor poorer. Victorian society is now remembered as two distinct ways of life. The areas around the docks and factories are typified by the East End of London. This was the haunt of the murderers and body snatchers Burke and Hare, and later, Jack the Ripper. The other impression of Victorian life revolves around sensitive ladies and gentlemen who held garden parties and lived in well designed houses in less populated areas. Newspapers carried letters reprimanding the poor for complaining of hunger and homelessness. One writer of the day suggested of the poor that they 'render no useful service', 'create no wealth', 'degrade whatever they touch' and 'are incapable of improvement'. Charities were set up by the wealthy either as a humanitarian exercise or to appease the church. The police put down riots by force. People died as a result of squalid, miserable living conditions.

Government intervention in the form of Public Health Acts (1848 and 1875) was a result of the realization by those responsible that health is tied to the environment in which people live. Properties rented to working classes were often leased by the owner to an individual who re-leased it to another, who let the building to a landlord who separated it into rooms and let them. The revenue from the dwellings was minimal, so landlords could not afford to properly maintain the properties. Any new housing stock would therefore represent an outlay with little income since wages were so low that people did not have the ability to pay even a reasonable rent. It became clear that poor housing was responsible for the health and wellbeing of many city dwellers.

Social and medical changes, which may have been instigated by the intolerable conditions suffered during the industrial revolution, resulted in better health care. More children survived to be adults, and adults lived longer. As a result the population began to grow rapidly. The population of many western industrial cities doubled every 25–30 years from 1800 to 1940.

The Working Class Dwellings Act 1890 was the first to direct its energies towards replacing the dilapidated buildings. Modern slums are the result of poor housing policy from the end of the Second World War (1945) to the mid-1970s. The high-rise, high-density local authority housing policy of that time was, in theory, an attractive proposition. Many of the early occupants were happy with their accommodation. The internal accommodation was generally of similar size or larger than the property the tenants were moved from. Unfortunately, some of the designs were substandard which has lead to the misconception that all high-rise accommodation is inferior. Construction costs per unit were lower. This was primarily because of the use of prefabricated concrete panels rather than the brick-cavity-block external walls used in private multistorey developments. Many of the construction processes used were unsuitable for the required purpose. Often only minimal insulation was used which resulted in rapid internal temperature changes. They only receive natural light on one, or at best two elevations. They are often damp because:

- they are difficult to ventilate
- the fabric of the building is not totally impermeable.

Often, because of the time involved in leaving the building, residents stay indoors considerably more than they otherwise would. Serious condemnation of high-rise buildings comes from health officials. There is now firm evidence that the mental health of the residents often deteriorates leading to apathy or more serious illnesses. Many of the buildings are vandalized and the areas become depressing to the people who live there. The public costs involved with health care and the constant remedial work required because of the inappropriate construction processes has persuaded some local authorities that high-rise blocks are a long-term liability. Many buildings less than forty years old have now been demolished.

Slum housing is not confined to any particular country. It is unfortunate that it is one of the few things common to nearly every country in the world. In Britain, as has been described, housing has been improved steadily so that whether those described earlier are slums is arguable. It may be simply a matter of comparison, after all, slum is a word defined by such relative terms as **overcrowded** and **poor**. The facts are though that very low quality housing is a threat to public health locally. It is often associated with pollution from local industry and from domestic properties as a result of inadequate or unrepaired sanitation. The buildings are seldom maintained properly, resulting in dampness and the resultant illnesses. Evidence for the social degradation associated with such habitation is often reflected in the crime figures.

1.2.6 New towns and cities

The Working Classes Dwellings Act of 1890 instigated the replacement of inner city slums with better local authority houses. However, the new properties were designed to give more space to the tenants so a land shortage was created in cities. Attempts to depopulate overcrowded cites began with the 1909 act dealing with housing and town planning. The original new towns were called Garden Cities. The construction of the first, Letchworth, began five years earlier in 1904. Welwyn Garden City was also started before the act was introduced. The designers took the now accepted view that people would prefer to live in an area which resembles the natural environment. This is difficult to achieve in inner cities. The cost of planting trees in central London would be dwarfed by the cost of the land they were to be planted on.

The New Towns Act of 1946 incorporated the garden cities and instigated the construction of Stevenage, Crawley and Harlow in England and East Kilbride in Scotland. The initial approach was to produce a complete plan of the settlement and then build to it. The suggestion that they were entirely new is not quite correct. In the main they were based around villages or small towns. Crawley for instance took in the small town of the same name and the villages of Pound Hill, Ifield and Three Bridges.

The latest of the new towns is Milton Keynes which took in the existing towns of Bletchley, Stony Stratford and Wolverton, and thirteen villages (including Milton Keynes). Its inception in 1978 was the largest of its kind in the UK. A priority in those early days was to attract builders. This was done by managing a council house waiting list of only a few weeks for those who worked on the new town. Those areas which were allocated to local authority housing were the first to be built on. These were followed by the lower priced private

houses and estates where expensive and low cost were mixed.

New towns and cities have been successful in some ways and unsuccessful in others. There is little doubt that overspill towns have had a positive influence on housing and public health. The conditions that many working class families lived in the 1930s and 1940s would today be considered unacceptable although they do still exist. Many of the slums of Glasgow, Manchester and London have been demolished and the people moved to modern properties. In the past new towns have been criticized for their lack of community spirit and bland architecture, especially the local authority housing. Milton Keynes Development Corporation has produced innovative buildings and attempted to form small communities on each estate. In some of the new towns the local authority properties are were often constructed in terraces which seem almost endless. In Milton Keynes the policy was to build only low rise buildings and plant large numbers of trees thus giving the impression of a smaller community.

Working on the design and planning of a new development such as a new town or city is perhaps a thankless task. It seems that no matter what changes are made in an attempt to overcome the problems learned from the past, new tenants are often unhappy with their new homes and surroundings. There is little doubt that mistakes are made. Inappropriate materials are sometimes used, and occasionally new construction methods are inferior to, or deemed to be less acceptable than older methods. There was, for instance, a rapid increase in the number of timber framed houses constructed in the 1970s and 1980s. A spate of bad publicity, the result of faults resulting from a lack of experience, and an apparent distrust of new systems, resulted in an equally rapid decline in timber framed construction. The apparent disenchantment of some of the early new town population may be social. Planners cannot force people to mix and form the sort of communities they came from, although the layout of estates is considered to influence feelings of isolation. The pre-new town people of the developed area – the locals – may feel a great intrusion by the new and numerically dominant population from other parts of the country. When the locals show their disenchantment the newcomers, no doubt, feel unwelcome, which adds to their own feelings of not belonging. However, new town development has been the primary example of improvements in housing conditions and public health during the twentieth century.

1.3 PLANNING

Prior to the Second World War attempts at government control over the development of land were minimal. During the war three official reports were published on the distribution of population and industry, land use planning and the protection of agricultural land. Subsequently, many of the proposals and recommendations were incorporated in the Town and Country Planning Act 1947, which effectively placed responsibility of all planning and development firmly with the national government in the guise of its new Ministry of Town and Country Planning. Local planning authorities were set up with the primary task of producing development plans for their areas. The important feature of the government act was to produce a twenty year plan which showed current and intended land use. In this way industry, commerce, domestic and leisure use was zoned and therefore segregated. For the first time any construction proposal had first to be submitted to the local authority for planning permission. The priority at that time was to check that the type of building was in accordance with the land designation. Further rights for the local authorities included the power to acquire sites, properties and land under a compulsory purchase order. It was the local authority that decided on a fair compensation. This has since been used to the previous owners' advantage and disadvantage.

The success or failure of planning legislation is, in the main, a reflection of the ability and/or desire of the local authority to implement national legislation. Legislation of this sort is often open to interpretation so that the use of planning strategy could vary depending upon the size of the local authority. In 1974 the Local Government Act created new governing counties, Avon and Humberside (now disbanded), and added new metropolitan districts such as Tyne and Wear, Greater Manchester and South Yorkshire.

1.4 PUBLIC DEMAND

The original function of the building envelope was to isolate the inhabitants from the worst aspects of the natural environment. As building technology has advanced and population densities increased the isolating function has broadened to include almost every possible outside influence.

1.4.1 Intruders

We often see ancient families represented as cave dwellers who light fires at the entrance to keep out bears and wolves. In the Middle Ages castles were constructed with moats and battlements to protect the inhabitants from marauding enemies. Modern buildings have sophisticated locks, garden fences or hedges, and the occasional box on an external wall advertising the installation of an intruder alarm system.

1.4.2 The weather

Once secure from the wolves, Saxons and thieves, the comfort of our homes and offices is determined by the extent of our isolation from the extremes of weather.

The British Isles is in a temperate region which reduces the incidence of heat waves, and is influenced by the Gulf Stream, so rarely experiences truly severe low temperatures.

1.4.2.1 High winds

The structure of buildings is expected to withstand winds of greater velocity than are normally experienced. If the property is to be constructed in a particularly windy region the design should not include extensive flat walls or other structures likely to be damaged by wind. If winds come from a dominant direction the pressure difference between the front and back of the building will, in high winds, produce suction loading which, in extreme cases, can pull out windows and doors and remove cladding. The pressure difference will also greatly increase the likelihood of draughts so openings should, if possible, be placed away from the direct current of air. Roof design is especially important in areas of high wind. Roofs with pitches from 15–30 degrees suffer from wind suction on the stoss side (Figure 1.3). News film of wind damage in progress often shows roof material flying suddenly skyward before other damage is done.

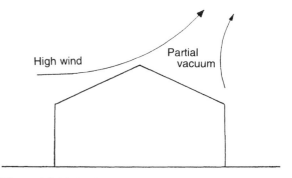

Figure 1.3 High wind across a low pitched roof

Figure 1.4 Frost action on a clay brick wall

The use of trees and other features as a wind break can be beneficial.

1.4.2.2 Frost

The expansion of damp materials when frozen is one of the most common causes of decay in brick walls and concrete structures (Figure 1.4). Water is absorbed so that the surface is saturated. When the water expands on freezing, part of the surface of the structure is destroyed. Pipes in the ground should be buried sufficiently deep to be permanently safe from attack by frost. Well-drained soils, such as sand or gravel, will not freeze as readily as for instance a clay soil which tends to retain moisture. Ground heave from frost action causes surface coverings, such as paving or concrete to lift. It can also damage foundations and pipes if they are too close to the surface.

1.4.2.3 Structural dampness

Moisture inside a building is not only a cause for discomfort, but promotes the growth of algae and moulds. Leaky buildings cause extreme discomfort and anxiety as well as internal damage, so the external envelope is reinforced by the addition of damp-proof courses, waterproof membranes and a variety of flashings and soakers. In cases of inadequate design, poor or negligent building practice, or lack of maintenance or misuse, houses and their occupants can suffer the effects of dampness. Prolonged dampness is believed to contribute to rheumatism and arthritis and respiratory illnesses such as bronchitis and pneumonia. Leaks are usually the result of a failure in a flashing, or roof tile or slate. Such problems must be remedied quickly because of the quantity of water entering the property.

Penetrating dampness moves horizontally through the external wall either because material has blocked the cavity or because it is a solid wall which for some reason has become permeable. Rising damp is the result of the failure or lack of a damp-proof course or membrane.

1.4.2.4 Natural light

A building without natural light can be a miserable place to live or work in. Direct sunlight in Britain is usually considered to have a positive influence on people. Properties with inadequate natural lighting are often associated with poor quality housing typified by slums. House agents consider descriptions such as 'sunny aspect' or a 'south facing garden' likely to increase the value of a property, whereas if the kitchen window faced into an alley they are unlikely to mention it.

1.4.2.5 External space

Residential density is one of the most important factors governing the quality of home life and is therefore a major consideration in project design. Homes which are well spaced and planned to fit in with the natural environment unquestionably provide better living conditions than those which have been planned with no consideration of the effects of living in a totally artificial environment. The high-rise, high-density local authority housing described earlier in this chapter was, in theory, an attractive proposition. The accommodation was generally of similar size or larger than the residents were used to. A further feature considered a benefit at the planning stage was that the buildings required less land per person and, given the same number of people per acre, provided large areas of open space which were not available in the back-to-back terraces which they replaced. Unfortunately, much of this open space was surfaced with concrete since grass could not withstand the trampling.

Given the above it is clear that the designers of high-rise accommodation were fighting against the natural environment rather than working, wherever possible, with it. The results were a social disaster which caused local authorities extra expense which has amounted to millions of pounds.

High-rise, high-density housing is not considered suitable for family accommodation especially where the very young or very old are involved.

Living space is discussed again in Chapter 6, The internal environment.

1.4.2.6 Extreme heat

Extreme heat is not common in the British Isles. On those unusual occasions when the temperature exceeds 30°C it is usually considered a benefit or at worst a moderate discomfort since we accept that it is a short-lived phenomena. However, if the climate is warming it is probable that service engineers will be installing increasing numbers of air conditioning units.

1.4.2.7 Orientation

The relationship between orientation of a building and environmental factors such as wind direction, available sunlight and ambient temperature is critical. To capture the sunlight so that it increases the feel-good factor, saves energy by increasing the internal air temperature, and allows for natural lighting in rooms where it is needed, for example kitchen and lounge, is an obvious advantage to the majority of the residents of the UK. In the earth's warmer regions the opposite is often the case. A room may be designed which has a small north facing window away from sunlight and wind action to where the occupants may retire when the weather gets too hot.

1.5 PROJECT DESIGN

If we are to build good quality homes members of the project design team should be able to show a commitment to:

- sharing knowledge
- feedback and evaluation
- environmental protection and improvement
- energy conscious design
- waste minimalization
- water conservation
- pollution prevention
- health and safety
- an environmental purchasing policy
- heritage conservation
- renewable energy

From *Environmental Code of Practice for Buildings*. BSRIA, (1994).

If environmental matters are not dealt with before and during the feasibility stage of a project, later alterations to the brief will cost money and cause annoyance.

If the design team are entrusted with the production of environmentally sound practices it is important that they are capable and sufficiently well educated in environmental issues. This often produces a discussion on the degree of environmental protection required. Take for instance a road scheme. Many environmentalists would expect some endangered plants and animals to be moved and some time dedicated to moving animals away from the area at a very early stage. Other environmentalists would hold up a road scheme until the tadpoles have matured to frogs, left the pond and made their way to safety. Another group would insist that all road building should be stopped since it destroys natural landscape and environments.

At the feasibility stage an assessment must be made of the effects on the environment of construction projects and vice-versa.

Case study – Bhopal

On the 3 December 1984 at Bhopal, India, the Union Carbide chemical plant exploded. Water was added to the methyl isocyanate (MIC). An exothermic chemical reaction occurred which increased the temperature and pressure in the pipes. There were two pieces of equipment designed to prevent catastrophic release of poisonous gasses. One was switched off and the other was in poor condition. As a result a pipe ruptured and twenty-five tonnes of MIC leaked from a storage tank and spread in a huge cloud over the boundary fence to the adjacent town. According to the official report, the number of deaths was 2153, though others have suggested that it was closer to 10000. The vast majority of those killed and injured lived in the shanty town which had developed around the perimeter of the plant.

There are always lessons to be learned from disasters.

1 Other companies produce the MIC as they need it so that only a few kilograms exist in the production pipeline, instead of the forty tonnes Union Carbide stored at Bhopal.
2 Domestic and industrial buildings should be kept separate, especially when the storage of hazardous materials is involved.

It is probable that if permission for the plant were refused on the grounds of storage of hazardous materials, Union Carbide would have used the Mitsubishi method of producing the MIC as and when it was needed. There is no doubt that accidents will happen, one of the many functions of those involved in planning is to minimize the incident and casualties when something goes wrong. Union Carbide has been generally considered at fault over this issue, but the question of the lack of planning control in India must also be considered.

1.6 CONCLUSION

We depend on our buildings to protect us from the worst aspects of the natural environment. However, history has shown that we can go too far. If people are housed in buildings which have little in common with the natural environment, or if we allow construction to proceed outside of local or national regulation, the results inevitably produce poor housing.

Evidence that planning is not a modern concept is found in the ruins of ancient civilizations. The Romans constructed their town roads in a grid system (a method copied by Milton Keynes Development Corporation) and built amphitheatres for public meetings. The Egyptians constructed great pyramids and the Greeks their temples. Many hundreds of years later, after suffering plagues, which killed a large proportion of the population, fires which destroyed property and life, and appalling public health problems following unplanned urbanization, Britain was awakened to the idea of a national and local planning policy.

2

Landscapes and soils

The aim of this chapter is to discuss at an *introductory* level, the earth sciences which are of significance to the construction industry.

Note: this is a very brief introduction to earth science. Students not going on to study geotechnics or similar, who wish to learn more on this subject should study any GCSE or A-level geology or physical geography text book.

2.1 INTRODUCTION

When competition for contracts intensifies, profit margins are reduced. In the 1950s and 1960s construction companies would expect to make a profit of around 20–30 per cent on the contract price. In times of recession profits of 3 per cent are more the norm. A lack of understanding of this branch of environmental studies has often resulted in high costs for construction companies.

While there is no intention to suggest that all construction students should become expert geologists, our understanding of the surface of the earth and how its different components will react to the loads imposed by the construction of buildings, roads or bridges is fundamental to construction technicians and professionals.

The construction of buildings in unsuitable ground is obviously not a new problem. In the seventeenth century BC a Babylonian ruler introduced the first building regulations which concerned the construction of buildings on unstable ground.

The code of Hammurabi (after Whitlow 1983)

If a contractor builds a house for a man, this man shall give the contractor two shekels of silver per ser (unit of weight) as recompense.

If a contractor builds a house for a man and does not build it strong enough, and the house which he builds collapses and causes the death of the house owner, then the contractor shall be put to death.

If it causes the death of the son of the owner, then the son of the contractor shall be put to death.

If it causes the death of a slave of the owner, then he (the contractor) shall give the owner a slave of equal value.

If it destroys property, he (the contractor) shall replace what has been destroyed, and because he did not build the house strong enough and it collapsed, he shall rebuild the house at his own expense.

If a contractor builds a house for a man and does not build it so that it stands ordinary wear and a wall collapses, then he shall reinforce the wall at his own expense.

In Britain today such punishments are not legal. What happens instead is a matter for the courts.

There are, however, strict contractural rules concerning the completion of projects on time, so that any unexpected hold-ups such as discovering that the foundations on the design are unsuitable for the soil, can cause the job to run over. Other financial penalties are imposed if, within some predetermined time, faults in the structure appear. In some cases the problems are so costly that the company is ruined. In many such cases disasters are avoided by a clear understanding of the surface of the earth. This study is known to the construction industry as geotechnics.

2.2 ROCKS

Rocks are made up of a mixture of minerals and represent the solid part of the earth's crust. They are separated into three groups each of which is classified by the mode of formation:

1 **Igneous rocks** are formed when magma cools and solidifies. Where this process occurs below the surface they are called intrusive and if the cooling process is slow the minerals form into large crystals. Granite is an intrusive igneous rock with large crystals of feldspar, mica and quartz. Extrusive igneous rocks are those which solidify outside of the magma chamber. In air or water they cool quickly not allowing time for the formation of large crystals. Some have a smooth glassy appearance, the surface of others exhibits the evidence of the escape of gases.
2 **Sedimentary rocks** are formed from sediments collected as a result of the erosion or weathering of existing rocks. Over time the sediment is consolidated. The rock is often named after the dominant sediment from which it was formed, hence siltstone, mudstone and sandstone. Others are named by appearance, a conglomerate contains different sized particle up to the size of pebbles. Limestones are predominantly deposits of calcium carbonate, often in the form of fossils. Chalk (a limestone) consists of vast quantities of microscopic fossils.
3 **Metamorphic rocks** are those which have been altered after deposition by changes in pressure and/ or temperature. The heating of sandstone by the movement of magma towards the surface may result in the production of quartzite. Marble is similarly formed from limestone, and slate under pressure from mudstone.

2.3 GEOLOGICAL TIME IN BRITAIN

The British Isles boasts an almost continuous geological history of the earth. It follows then that strata of almost every age supports construction projects. In general the oldest rocks can be found to the west of Britain, notably Cornwall and the Western Isles of Scotland, and the youngest in the east where the sedimentary deposits of southeast England and East Anglia are found.

Table 2.1 shows how geological time is separated into eras, periods, epochs and stages. The eras refer to the antiquity of the animals as they seem to us in their fossil form. The suffix -zoic simply means of animals, while the prefixes Eo-, Palaeo-, Meso- and Ceano- mean dawn, old, middle and new. The periods are generally named after the location where they were first discovered, or some outstanding characteristic.

Pre-cambrian rocks are limited to occasional outcrops in Cornwall and Leicestershire (Charnwood Forest) and are common off the west coast of Scotland. The best examples are the metamorphic and igneous rocks of the Isle of Lewis.

Cambrian (began 570 million years ago) rocks are predominantly Welsh sedimentary rocks such as limestones and mudstones. There are also metamorphic rocks such as Welsh slate which has the reputation of being the finest in Europe. The Isle of Man is almost entirely of Cambrian age. Typical British Cambrian fossils include trilobites, algae and graptolites.

The **Ordovician** period (500 Ma) produced a similar suite of rocks which outcrop in Wales and in patches along the west coast of Britain. The fossils include trilobites, brachiopods and simple molluscs.

During the last two periods mountains were being slowly forced up in northern England, Wales and Scotland.

During the **Silurian** (440 Ma) the same sort of rocks were forming around the Welsh border as the Caledonian mountain building came to an end.

The **Devonian** period is special because for the first time there was sufficient oxygen in the atmosphere to oxidize (rust) the metallic minerals in rocks. One of the most distinctive rocks of any period is the Old Red sandstone. Other Devonian deposits include shales, limestones and slates. The land was colonized by plants and in the sea the first cartilaginous fish, the early sharks, appeared.

The **Carboniferous** period (345 Ma) was named after the coal seams found in the shales and sandstones. A common carboniferous sedimentary rock is Millstone Grit, a very coarse sandstone, found in Yorkshire, Lancashire, parts of North Wales and much of Devon. Another is the Carboniferous limestones which were deposited in warm seas. The variety in the deposits indicates constantly changing depositional environments during the carboniferous period.

The **Permian** (280 Ma) was typified by desert conditions and by the lifting of the land to form the Pennines and other smaller ranges of hills. Permian rocks are found in a thin band stretching north from Nottingham to North Yorkshire. The Permian denotes the end of the Palaeozoic era and the beginning of the Mesozoic. This boundary, sometimes known as the

Table 2.1 Geological time

Era	Period	Epoch	Event(s)
Cenozoic	Quaternary	Holocene	Civilizations
		Pleistocene	Great Ice Age
	Tertiary	Pliocene	Early hominids Primates
		Miocene	Alpine orogeny African plate moving north
		Oligocene	Sediments deposited in synclines
		Eocene	London clay African plate collides with Euro–Asian
		Palaeocene	Volcanoes around Scotland Mammals become dominant on land
Mesozoic	Cretaceous		Relative fall in sea level Chalk landscapes
	Jurassic		Reptiles dominate land sea and air
	Triassic		Sea dominated by ammonites
Palaeozoic	Permian		Great extinction Hercynian orogeny
	Carboniferous		Coal measures Fluctuations in sea level
	Devonian		First fish and insects
	Silurian		Caledonian Orogeny
	Ordovician		Trilobite brachiopods and molluscs dominate oceans
	Cambrian		Planktonic algae Invertebrates
Eozoic	Precambrian		Blue green algae Volcanoes in England and Wales

Permo-triassic, is represented by a mass extinction. Trilobites disappeared from the fossil record and brachiopods numbers were reduced dramatically.

The **Triassic** period (225 Ma) was notable for the development of reptiles on land. In the seas ammonites replaced trilobites, and bivalves occupied the niches vacated by brachiopods. On land the lizards evolved from the amphibians. It is generally thought that the first mammal evolved from the early lizards towards the end of this period.

The rocks of the Triassic include the marls, sandstones and Bunter Pebble beds and are found in areas of England, notably in a broad band from Somerset to Northumberland.

The Jurassic (190 Ma) is most famous for its dinosaurs. However it was not the only period in which they lived. In the Lower Jurassic the alternating layers of limestone shale and clays are known as Lias (a derivation of the Cornish word meaning layers) (Figure 2.1). These sedimentary deposits are evident in a broad band from the Dorset to the Yorkshire coasts.

The common fossils are marine, especially ammonites and belemnites. The dominant plants were

Figure 2.1 Blue Lias, Lyme Regis

conifers and soft tissue plants such as giant ferns and cycads. The Upper Jurassic deposits include Oxford Clay from which many of our bricks are made, and the excellent Portland Limestone which was transported to London in great quantities for construction use in the nineteenth and early twentieth centuries.

The **Cretaceous** period (136 Ma) saw the development of the early mammals and birds, but was still dominated by the reptiles, notably Tyrranosaurus rex and Triceratops. Flowering plants, which includes broad leaved trees, began to take over from giant ferns and cycads. The early Cretaceous deposits are the greensands and clays of southeast England. The common Upper Cretaceous rock is the chalk which forms the Downs of Southern England. The Mesozoic, like the Palaeozoic was terminated by a mass extinction. The Cretaceous–Tertiary boundary is famous for the apparently sudden disappearance of dinosaurs, ammonites and many other plants and animals.

The **Tertiary** period (65 Ma) is most famous for the Alpine mountain building episode. In southern England this resulted in great folds, evidence of which are the North and South Downs with the Weald Valley between. With one exception the deposits are all sands in southeast England. The exception is the London clay which was the material used by the original London Brick Company.

The **Quaternary** period (2 Ma) is characterized by the Great Ice Age. It is further divided into stages

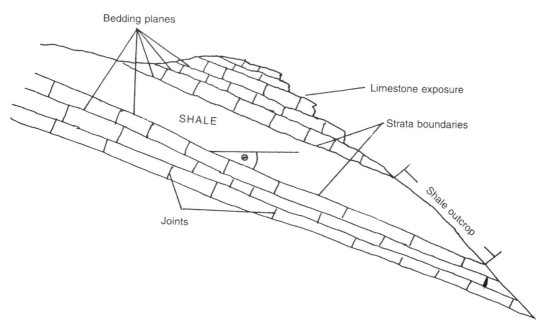

Figure 2.2 Stratigraphic terms

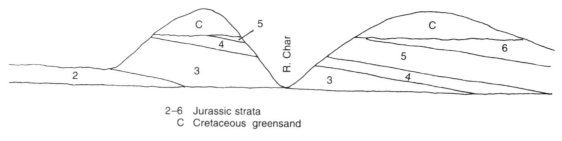

2–6   Jurassic strata
C   Cretaceous greensand

Vertical scale exaggerated, about 3 miles horizontally

Figure 2.3 Unconformities between Jurassic and Cretaceous strata

which indicate either a warm or cold climate. At least three of the cold stages were severe enough for ice to cover much of Britain. The deposits are therefore mostly either sands and gravels, glacial till or occasionally peat and other organic soil deposits. The dominant land animal is *Homo sapiens*.

The last two periods are confused by the introduction of epochs, five in the Tertiary and two in the Quaternary. There is a school of thought which believes that the use of Quaternary and Tertiary is not acceptable and that the epoch names should be used instead. For the moment it is safer to suggest that the Pleistocene is the glacial epoch of the Quaternary Period, and the Holocene the post glacial (see Table 2.1).

2.4 STRATIGRAPHY

To the geologist stratigraphy is the study of the succession of distinct layers of rock or deposit in terms of geological time. The basic rule is the 'law of superposition'. This proposes that each new deposit forms on top of an older one. A continuous deposit of a particular rock or sediment is the bed. A bedding plane represents a break in sedimentation and is followed by a further deposition of either the same material or a different deposit (Figure 2.2). A bedding plane therefore can have the same material above and below it. Where a bedding plane separates different beds it is a strata boundary. Vertical cracks are called joints and may be filled with finer sediment.

Note that an outcrop is not necessarily visible at the surface. It may be covered by soil. In the hypothetical example in Figure 2.2 the top of the hill is a limestone outcrop. The part of the outcrop which is visible is the exposure. Where the bed is tilted the direction and the angle measured down from the horizontal is the dip.

Obviously, deposition is not the only mechanism at work. For long periods erosion of existing rock

occurs. This may be the result of many processes but especially by weathering, river or glacial action. The result is that part of the stratigraphic sequence is removed and a much later deposition appears to follow immediately. This is known as an **uncomformity** (Figure 2.3). If the earlier beds dipped the unconformity will be angular, if not it is parallel.

The most common method of stratigraphic dating is by comparison of available fossil data (Figure 2.4). Many beds are given the name of the most common fossil. Another method is to compare the deposit sequence of beds but care in required here since the very similar deposits may be produced in different geological periods.

Borehole 1   Borehole 2

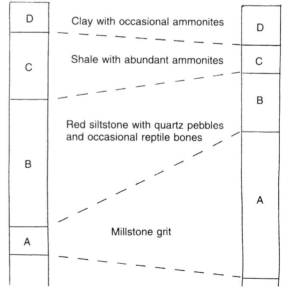

Figure 2.4 Stratigraphic correlation of two sites

Question 2.1

Refer to the section on earth history and Figure 2.4.

1 To which period does horizon A belong?
2 What can be said about the age of horizon B?

Answers

1 Carboniferous
2 It is Mesozoic

Almost every aspect of geology has some application in the construction industry. For instance, the raw materials used in the production of cement are clay and chalk. Many of the countries cement works are situated on, or close to Upper Jurassic and Upper Cretaceous deposits. The countries biggest brick manufacturers are to be found by Upper Jurassic deposits, especially the Oxford clays.

Question 2.2

Refer to Figure 2.5 and fill in the gaps in the sentences below.

1 The Ordovician limestone _____ at A.
2 The near horizontal lines in the limestone represent _____ _____.
3 The near vertical lines in the limestone are _____.
4 The outcrop of Devonian Conglomerate is indicated by the _____.
5 E represents the _____ _____ between the Ordovician limestone and the Ordovician sandstone.
6 The _____ at the upper boundary of the limestone is 10 degrees.
7 The Ordovician was followed by the _____ and not the Devonian.
8 The strata boundary between the Ordovician and Devonian deposits is therefore an _____.

Answers

1 outcrops
2 bedding planes
3 joints
4 exposure
5 strata boundary
6 dip
7 Silurian and unconformity

2.5 TECTONIC FEATURES

Previous sections have shown that one of the primary considerations concerning the stability of structures is the ability of the ground to support them. In small and medium sized buildings this is probably the only consideration. However for larger buildings and civil engineering works we must consider the stresses and strains which occur at or near the surface of the earth.

On the largest scale we have a considerable body of evidence which suggests that the earth's crust comprises a number of plates which move relative to each other. On a smaller scale, fault and folds create permanent and intermittent weakness in strata.

2.5.1 Faults

The most famous of all faults, and one of the largest, is the San Andreas Fault on the west coast of North and Central America (Figure 2.6). Although both plates are moving in the same direction, they have different velocities so that relatively speaking, one is passing the other at 50 mm year^{-1}. Unfortunately their passing is not a smooth continuous process. The edges of the plate move in a series of unpredictable jolts rather like sliding the tip of a finger across a desk. The elastic potential energy which builds up is released intermittently as an earthquake.

The early construction methods in the USA were similar to those used in Europe in the nineteenth century. Underground pipes were rigid and buildings of two or three stories were made of timber and brick.

Case study – The San Francisco earthquake and fire

Shortly after 0500 on the 18 April 1906 San Francisco experienced the first of three tremors the last of which measured 7.9 on the Richter scale. Construction methods were similar to those used in Europe and were, therefore not designed to withstand earthquakes. Many of the buildings collapsed fracturing water and gas mains. The gas added to the fires originally caused by electric cables shorting and stoves being overturned. Most of the buildings were detroyed by fire because fractured water mains left the fire service unable to cope. By the time the fire was put out three days later, 28 000 buildings had been destroyed and 450 people killed.

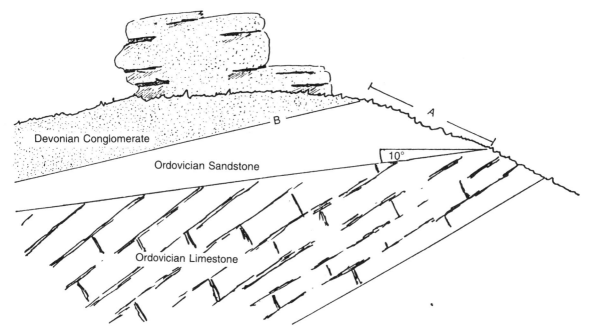

Figure 2.5 Palaeozoic strata

Figure 2.6 The San Andreas Fault

Since then construction methods have changed so that when, in 1989 an earthquake measuring 6.9 on the Richter scale shook San Francisco for 15 seconds, skyscrapers swayed by up to three metres but remained undamaged. Many of the sixty-seven fatalities were in their cars.

In other parts of the world, New Zealand for instance, a different sort of earthquake occurs. Here a pressure wave moves across the landscape and causes the shape of a building to change (Figure 2.7). The damage this causes to buildings is lessened by allowing the wall plate to move relative to the roof. When the wave front reaches the building the wooden wall tips into the room at the ceiling, then goes back to the vertical, then it tips out and finally back to the vertical again.

Smaller faults occur in almost every type of strata. A normal fault is indicated by a break or dislocation in the strata (Figure 2.8a) with the downthrow and

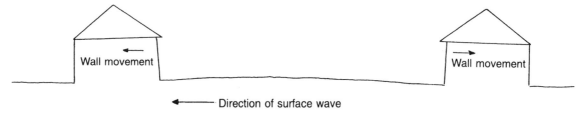

Figure 2.7 Single story house and pressure wave earthquake

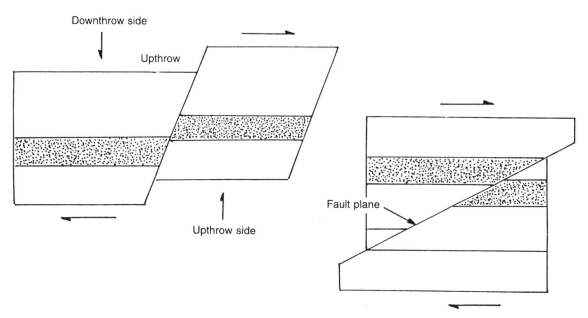

Figure 2.8 (a) Normal fault formed by tension; (b) thrust fault formed by compression

upthrow blocks usually quite evident. In the San Andraes Fault the relative movement is horizontal, in smaller faults it is often vertical but it may be at any angle and almost any dimension.

Faulting occurs when the strata is fractured so that the beds on one side appear higher than those on the other (Figure 2.8b). The most common, hence the use of the term normal fault, are those created by horizontal tension or by vertical compression. Thrust faults on the other hand are only formed by horizontal compression.

If the ages of the strata are known it is possible to ascertain the age of the faulting and the time for which it has been stable. In Figure 2.9 the fault has occurred in the Lower Jurassic rock but does not appear in the Upper Jurassic.

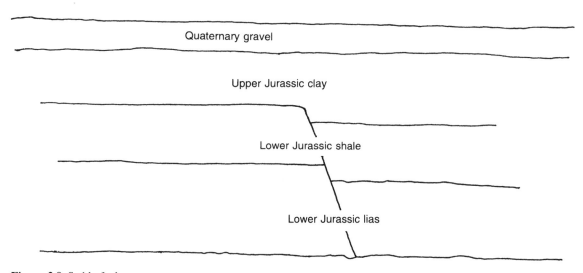

Figure 2.9 Stable fault

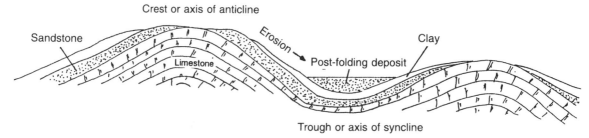

Figure 2.10 Folding

2.5.2 Anticlines and synclines

During the Tertiary period the African Plate moving north collided with the Euro–Asian plate. The evidence of the horizontal pressure is a series of waves in the landscape across Europe, the greatest of which is the Alps. The amplitude of the waves reduces with distance so that the most northerly evidence is found in the hills and valleys of Southern England. Figure 2.10 shows how the horizontal compressive forces can fold the strata into different shapes.

This is especially clear in the Isle of Purbeck where one can easily see the Purbeck Hills (Figure 2.11a). Evidence of the force from the south is also seen on a smaller scale at Stair Hole where the Jurassic rocks have been folded into the Dorset Crumple (Figure 2.11.b).

2.6 RIVER VALLEYS

Much of the lowland landscape is dominated by river valleys or drainage basins. A line representing the boundary of the basin, the watershed, follows the high ground which separates drainage basins (Figure 2.12). The rainfall which falls onto the catchment either evaporates from the surface or through plants, or makes its way through the soil into the rivers and to the sea. All countryside is therefore mapable as an arrangement of drainage basins.

2.6.1 Development of fluvial systems

Fluvial, or river valley development may be divided into two sections. The first is concerned with the development of individual river valleys, and the second the development of drainage systems. Obviously the two develop synchronously but they are treated as separated topics.

2.6.1.1 Stream initiation

The great variety of surfaces upon which fluvial systems have developed makes generalizations about their initiation unacceptable. The development of streams on an ameliorating (warming) periglacial landscape has little in common with fluvial initiation resulting from tectonic activity or a falling sea level except that they are all a function of water acted on by gravity.

An early model (Horton, 1945) suggests that from a flat surface on high ground overland flow could occur. Some authorities consider this unlikely except on a semi-arid poorly vegetated landscape, but if there were no rivers in the area at that time the water table would rise until the ground was saturated. Bogs on flat high ground are quite common. Once the ground is saturated rills or small streams would develop. Because of minor differences in resistance one rill or stream will become slightly deeper than the others. Then, being closer to the water table, it will be the last one to stop flowing in drying weather. It would therefore erode for longer and become deeper and wider and actually engulf the other rills rather than capture their water.

The development of drainage patterns depends largely on the geology of the area. If the area were flat and dipping only slightly a series of parallel streams would evolve producing a trellis-like pattern, if the area dipped towards a point a dendritic pattern would evolve (Figure 2.13).

The drainage basin is an open system which forms part of the water cycle. The primary input is precipitation in the form of rain, snow and hail, the secondary is insolation (the light which reaches the surface of the earth). Outputs are the loss of water to the sea, and through evapotranspiration. To a lesser extent rivers also form an important part of the rock cycle, that is they erode and transport material to the sea, so we should include rocks and minerals in our

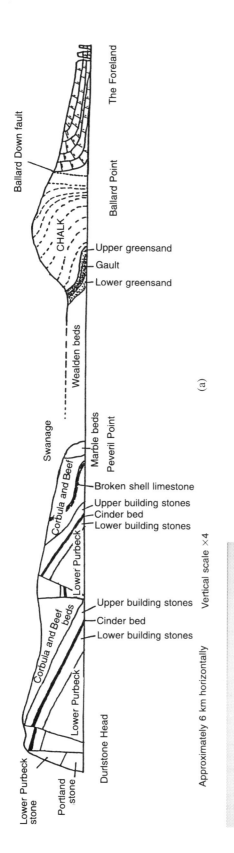

(a)

Ballard Down fault

The Foreland

Ballard Point

CHALK

Upper greensand
Gault
Lower greensand

Wealden beds

Swanage

Marble beds

Peveril Point

Corbula and Beef

Broken shell limestone

Upper building stones
Cinder bed
Lower building stones

Lower Purbeck

Upper building stones
Cinder bed
Lower building stones

Corbula and Beef beds

Lower Purbeck

Durlstone Head

Lower Purbeck stone

Portland stone

Vertical scale ×4

Approximately 6 km horizontally

(b)

Figure 2.11 Evidence of Tertiary folding (a) the Purbeck Hills – synclines and anticlines; (b) the Dorset Crumple

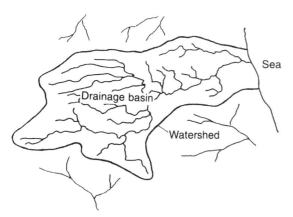

Figure 2.12 Drainage basin

list of inputs and outputs. Finally, river water does occasionally get contaminated, e.g. effluent, agricultural fertilizers, industrial spillages, which, if untreated, find their way to the sea.

2.6.1.2 Precipitation

Rainfall varies in intensity: a thunderstorm may produce more rainfall in an hour than is normally transported by the river in a day, but they usually

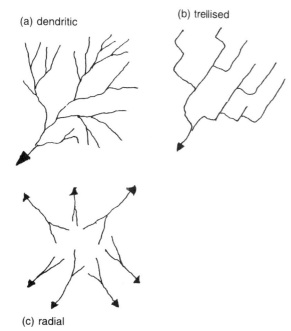

(a) dendritic

(b) trellised

(c) radial

Figure 2.13 Drainage patterns

follow warm dry periods and may be so localized that they do not affect the whole basin. The worst senario is a series of fronts associated with a low pressure system. These usually produce persistent heavy rain over large areas.

2.6.1.3 Channel shape

Channel shapes can be described numerically using the hydraulic radii. The hydraulic radius of a river channel is the ratio of the cross sectional area and the wetted perimeter.

Question 2.3

A stream has a rectangular cross section. It is 3 m wide and 1.5 m deep. What is its hydraulic radius?

Answer

Wetted perimeter $= 1.5 + 1.5 + 3.0$

$$= 6.0\,\text{m}$$

Cross sectional area $= 1.5 \times 3.0 = 4.5\ \text{m}^2$

$$\text{Hydraulic radius} = \frac{6.0}{4.5} = \frac{4}{3}$$

2.6.1.4 Velocity

In a symmetrical channel on a straight stretch of river the velocity increases where the friction is lowest (Figure 2.14). Lines joining places of equal velocity are called isovels. The velocity is therefore partly dependent upon the shape of the channel.

The roughness of the wetted perimeter also influences the velocity. In a boulder-strewn river channel the flow is restricted by the friction. As it flows towards the mouth the particles become smaller and

Isovels: $v_1 > v_2 > v_3$

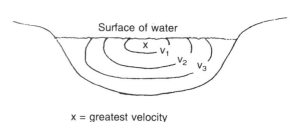

x = greatest velocity

Figure 2.14 Isovels

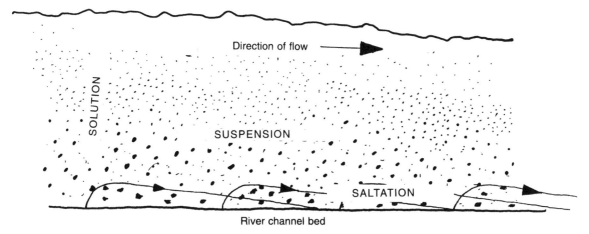

Figure 2.15 Transport mechanisms in rivers and streams

eventually cohesive. The most common misunder-
standing on this subject is the difference between the
velocity near the source and the mouth of a river. Near
the source the channel is steeper and more erosive.
There are more eddies, waterfalls and chutes but the
mean velocity is greater in the lower part of a river
channel where resistance is lower.

2.6.2 Transportation

Figure 2.15 shows three physical processes involved
in the transport of material downstream. They are
solution, suspension and bed-load. The latter is by
saltation or traction.

2.6.3 Erosion and deposition

In this sense erosion refers to the bed and not the
particles. If we consider three groups of particles,
small (clay and silt) medium (sand) and large
(gravel to boulders), we can easily imagine that the
velocity required to move boulders is greater than
that required to move sand. As the particles become
smaller less energy is required to move them.
However, the finest particles become cohesive and
erosion is reduced (Figure 2.16). In essence, large
boulders have inertia because of their mass, and
clay particles are difficult to move because they
form a solid accumulation.

The transportation of load also depends on velocity.
Once in the body of the water clay particles will
remain in suspension at very low velocities. As
particle size increases the velocity required to keep
them in suspension increases.

Three points worth remembering.

1 All particles require a greater velocity to lift them
than to maintain them in suspension.
2 The velocity referred to is local and not the mean
for that stretch of the river. In effect it is often the
turbulence which produces the greatest erosion of
the river bed.
3 The difference in velocity between erosion and
deposition decreases as particle size increases.

Deposition occurs when velocity and turbulence
are reduced. There are four common mechanisms:

1 as a river broadens the wetted perimeter increases
and the associated friction reduces the velocity
close to the channel surface.
2 as a river enters a sea or lake.
3 reduced discharge caused by dry weather.
4 when the turbulence is reduced locally.

2.6.4 Grade and longitudinal section

The longitudinal profile of a river may be represented
by a graph of altitude, on the y-axis, and distance from
base level (usually the sea) on x-axis. When the river
is in its early stages of development the longitudinal
section appears as a series of jumps between lakes and
ponds often where the resistance of outcropping rock
changes. As the channel ages outcrops of harder rock
will be eroded and become part of the river valley.
The upper reaches of a river channel are generally
erosional and the lower reaches depositional. Grade is
a state of near equilibrium. It is usually expressed as

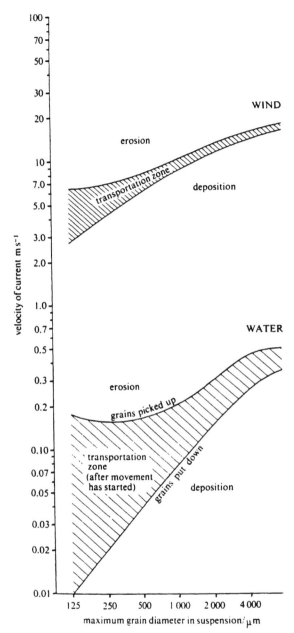

Figure 2.16 The effect of velocity on erosion and deposition (reproduced by kind permission of The Open University)

an event which occurs in a mature river valley when the long profile is smooth, regular and concave, and steeper at the source. It is then that the river ceases its vertical erosion and broad meanders are produced with time. Fluvial systems are, however, subject to constant external changes such as extreme rainfall,

river management schemes and slope erosion processes.

In the upper erosional zone and the whole length of young rivers, the landscape produced is usually typified by v-shaped valleys (Figure 2.17a). As the river approaches the sea the landscape is dominated by river terraces (Figure 2.17b).

2.6.5 Flood plains

When a river reaches grade it cannot erode vertically. Lateral erosion produces meanders which in turn produce oxbow lakes, point bars and pools (Figure 2.18). The velocity of the water is greater around the outside of the meander than the inside causing the water to pile up against the external side of the bend. This creates pressure difference which is dissipated by water and sediment moving across the channel producing erosion on the outside and deposition on the inside. The exception is when the river is in flood. Then erosion tends to occur right across the channel.

The energy of a river is at its greatest at the bankful stage. Any increase in depth will result in overflow onto the flood plain. The sudden increase in the wetted perimeter and friction slows the water and produces a thin layer of silt and clay over the plain.

2.6.6 Bluffs and terraces

During the evolution of a river valley rapid changes sometimes occur which may cause the long profile to change shape. These changes include river capture, especially where the captured river is comparatively large, climatic changes especially precipitation, or base level changes, such as a rise or fall in sea-level. A river is said to be rejuvenated when sea-level falls, since the velocity and therefore erosion will increase. The results of these changes are seen in the landscape as bluffs and terraces which are often named after the type site. Paired river terraces (Figure 2.19a) are slightly unusual because the meandering river often obliterates the terrace on one bank (Figure 2.19b). Much of the British landscape is the result of fluvial erosion and deposition.

2.7 GLACIAL LANDSCAPES AND SOILS

Much of Britain has been covered by ice on at least three occasions during the Pleistocene. Geologists believe that the earth has been through many more cold phases but their effect on the British landscape is

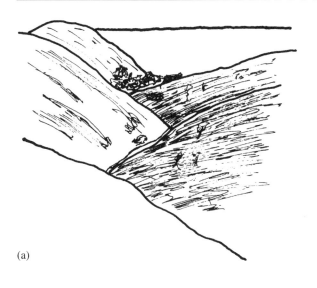

(a)

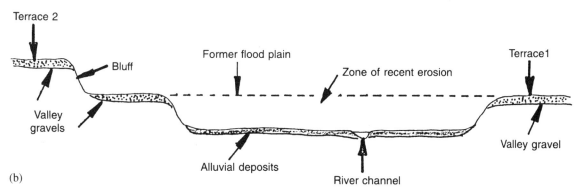

(b)

Figure 2.17 (a) Typical immature river in a hardrock outcrop; (b) typical lowland mature river valley dominated by deposition with occasional erosion

uncertain. The warm and cold stages of the Pleistocene were the product of global climatic change. In the last three cold stages great sheets of ice built up over much of Britain (Figure 2.20). Current theory suggests that only the south of England has always been free from ice.

Movement of ice in glaciers is primarily the result of the force created by the increasing weight of snow and ice in high altitudes. Ice is plastic at such a large scale which adds to the movement at the ice front. The proportion of a glacial base which slides on the surface has an effect on the landscape produced.

2.7.1 Glacial erosion

In effect the glacier acts as a giant sheet of abrasive paper. The force acting on the bed of a valley glacier is a function of its mass and centre of gravity. The continual freezing and melting as the pressure changes causes ice to grab or pluck at the surface. Rock debris is then held by the ice and provides the abrasion. As the abrasion continues the particles become finer producing at the base of a temperate glacier a layer of clay and silt with pebbles, cobbles and occasionally boulders.

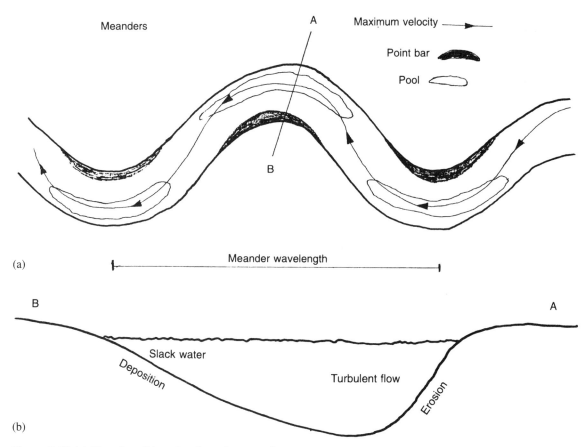

Figure 2.18 (a) Meanders; (b) section through a meander

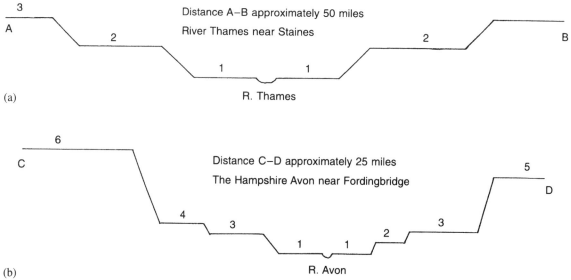

Figure 2.19 (a) Paired river terraces; (b) Terraces of the Lower Avon

Figure 2.20 Glacial Britain

2.7.2 Landscapes

Valley ice will typically produce an almost semi-circular or **U-shaped valley** (Figure 2.21) with a rock surface scratched or striated by the rocks in the ice.

Cirques (Figure 2.22) are formed when a hollow in the side of a mountain is filled with ice. The ice rotates slowly around a pivot point. The pressure at the **headwall** produces a steep cliff of angular rocks at its surface. This is the result of plucking and frost shattering. In complete contrast the base is smoother and flatter as a result of abrasion and increased pressure. At the rock lip the mass of ice is thinner, so the abrasion is reduced. Partial melting produces a small stream at the rock lip which, in time cuts a small v-shaped valley.

Roche moutonnees (Figure 2.23) occur when a glacier moves across a landscape of exposed rocks. The increase in pressure on the stoss, or up-valley slope, causes the ice to partially melt. The erosion is therefore less severe and produces a smooth slightly rounded surface. On the lee slope the pressure reduces and the ice re-freezes. Fragments of rock are plucked from the surface leaving a jagged down-valley slope.

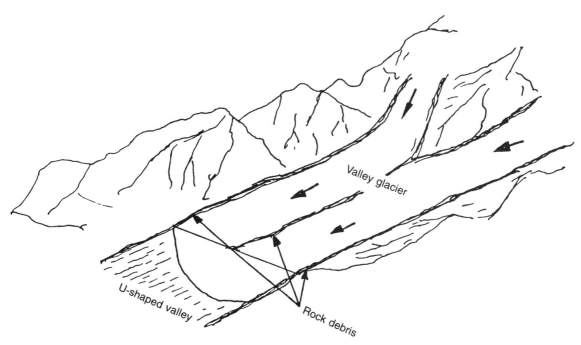

Figure 2.21 A U-shaped valley

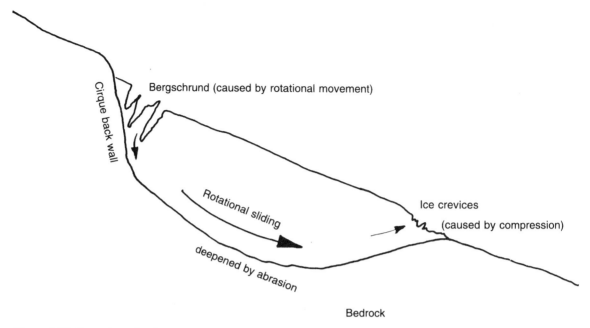

Figure 2.22 Formation of a cirque

Crag and tail (Figure 2.24) landscape is created by a similar process but on a greater scale. The major difference is that glacial drift is eroded from the stoss side and deposited in variable quantities on the lee side.

2.7.3 Deposition

The general term for all glacial deposits is **drift**. It is advisable, however, to refer specifically to gla-cial drift since some workers, especially geogra-phers, apply the term to any unconsolidated mate-rial at or near the surface. The particles of rock inside the ice can be of almost any size, from boulders to microscopic clay particles. It was, therefore originally called boulder clay. However, that name suggests that particles of those two sizes are always present in glacial deposits and, since this is not the case, the name has been dropped in favour of **till**.

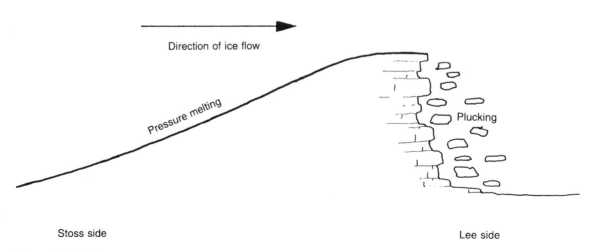

Figure 2.23 Roche moutonnee

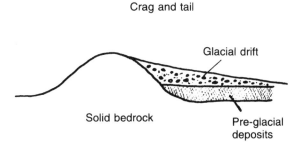

Figure 2.24 Crag and tail landscape

Tills are unsorted glacial deposits which contain all shapes and sizes of rock debris. The particles are generally sub-angular to angular. The lithological composition of till may explain a great deal about the origin and/or direction of a glacier. In Shropshire the till contains **erratics** from the Western Islands of Scotland. East Anglia and much of central England is covered by a chalky till with a clayey matrix. A very few contain erratics which are believed to have originated in Scandinavia.

Material carried by the glacier and deposited to produce changes to the landscape is **moraine** (Figure 2.25). Three examples are:

1 **Lateral moraine** Frost shattering occurs on the rocks above glaciers so that debris falls to the upper edges at the surface.
2 **Ground moraine** When the glacier melts *in situ* the debris produces a featureless till.
3 **Terminal moraine** When the debris moves to the snout it produces a ridge of till marking the final position of the glacier.

Drumlins are elongated mounds of till which form parallel to the direction of the glacier. They may be as much as 50 m high and 1000 m long and have the appearance of small elliptical hills. There is a good deal of disagreement about the mechanics of formation but since they form behind the terminal moraine they are either subglacial or postglacial deposits. The lack of sorting seems to indicate a subglacial deposition.

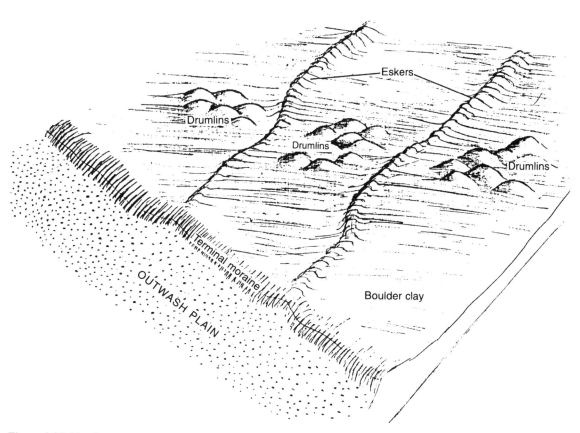

Figure 2.25 Moraine

Fluvioglacial deposits (often reworked tills) are created by glacial meltwater and may be in front of, or behind the terminal moraine. As the glacier melts, outwash sands and gravels are sorted by the movement of water to form fluvioglacial deposits. Examples include:

- **Kame terraces**   As the surface of the glacier melts streams form in troughs between the valley side and the glacier and wash the debris to the valley floor.
- **Varves**   Layered deposits of silt or clay found near glacial margins. They possess a distinctive banding created by the alternating deposits of organic free material in the winter and peaty deposits in the summer.
- **Eskers**   As the climate warms the base of the ice melts producing subglacial streams which deposit mounds of coarse sand and gravel. These can only occur when the ice is retreating.
- **Kettle holes**   As the fluvioglacial and ground moraine is deposited large blocks of ice are buried. When they melt they leave depressions which often fill with water.

2.8 LIMESTONE LANDSCAPES

Limestones are calcareous sedimentary rocks. They are deposited in warm shallow seas by the precipitation of calcium carbonate generally in the form of fossil animals of varying sizes. Their appearance and structure depends upon the type of precipitation. Some are formed by the fossilization *in situ* of fossils such as bivalves. Others form from much smaller fossils which are re-worked by sedimentary processes to form hard limestones such as those found in Yorkshire and on the Isle of Purbeck. Chalk is formed by the precipitation of microscopic fossils and forms a softer, less resistant and more porous rock than other limestones. The more resistant and denser limestones are excellent rocks for use in the construction industry as building materials.

Limestone landscapes are often riddled with evidence of the action of water because limestones readily weather by solution. Water percolates through most soils and limestone is no exception. However, when a limestone landscape rests on an impermeable strata (Figure 2.26) the resulting underground drainage can produce pot-holes and caverns.

A pipe is a hole often found in a limestone landscape, which has filled with topsoil. Great care should be taken at the initial survey stage since this sort of landscape is unsuitable for most construction purposes. This type of ground may also exhibit dry valleys, surface depressions and streams which disappear into the ground at swallow holes. While these do not in themselves constitute a problem the fact that the water has disappeared below ground does. Caves are formed as enlargements of the underground streams. Evidence of percolation is in the form of stalactites and stalagmites. The landscape above often exhibits flat exposures known as limestone pavements as found in Cumbria.

Typical underground drainage in limestone

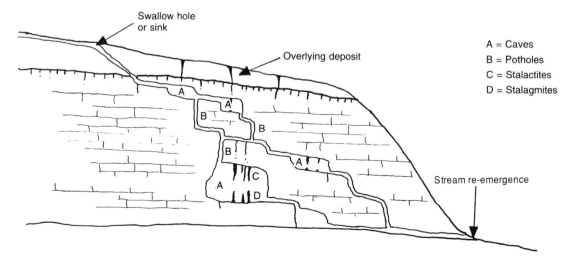

Figure 2.26 Limestone drainage

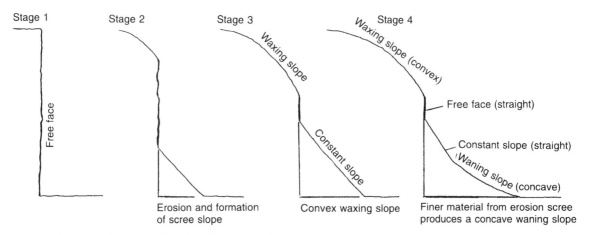

Figure 2.27 Formation of a typical slope from a free face

The soil immediately above limestones is usually a soft paste of weathered rock. Close to the surface the influence of other more resistant local rocks may dominate the soil type.

The result of weathering by solution on limestone landscapes is that voids of a variety of sizes and shapes are to be expected which may cause serious problems in the construction of foundations, reservoirs and tunnels. Construction on a limestone landscape should therefore always begin with a detailed and comprehensive ground investigation.

2.9 SLOPES

An understanding of the problems involved in slope stability is of considerable importance to the construction industry, especially civil engineering.

2.9.1 Slope form and evolution

The formation of a slope from a free face has four observable elements (Figure 2.27). Initially the **free face** (which is straight) erodes especially near the top where soil creep occurs producing a convex **waxing slope**. This material drops and forms the **constant slope** which forms at the angle of repose. As the slope develops material falling onto the constant slope from the waxing slope and the free face are weathered and move downslope to form a concave **waning slope**.

2.9.2 Slope stability

The downslope motion of the products of weathering is **mass movement**. A stable slope may be considered an open system in a steady state, that is, there is a constant input and output of material, while the angle

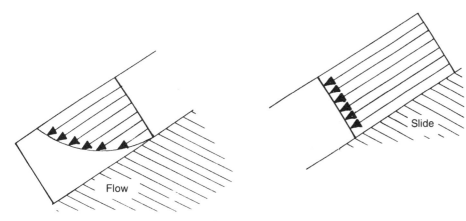

Figure 2.28 Mass movement – slide and flow

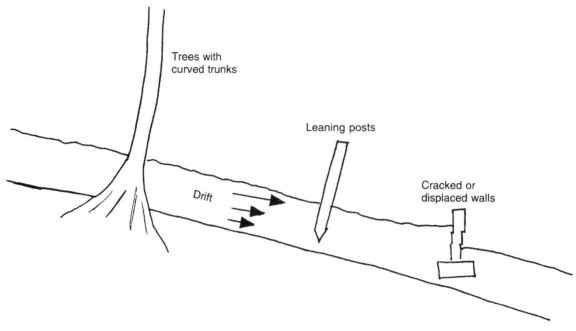

Figure 2.29 Evidence of soil movement

of repose is maintained. It is always a temptation for scientists and technologists to derive figures and formulae which produce statistically-based definitions of real life situations, and their usefulness is not in question. However, because of the variations in local conditions such as weather, lithology and vegetable cover, statistically-based definitions of slope evolution are often considered less dependable than the results of field work.

The descriptive classification of mass movement recognizes two main types. The catastrophic landslides, are the result of **slide**. That is the material slides across a plane, often as a sheet, with equal velocity throughout the mass (Figure 2.28).

Alternatively **flow** occurs when the velocity is greater at the surface and reduces to near zero at the plane.

The slow downhill movement of surface material is often called **soil creep**. The processes responsible are in themselves only capable of producing slight movement, but their variety and persistence is indicated by bulging walls, trees with curved trunks, leaning posts and turf rolls (Figure 2.29). Unstable or moving soil is also indicated by cracked, terraced or hummocky ground.

When the soil freezes particles lift at a right-angle to the slope. When it thaws it falls back under the influence of gravity. The quantity of material lifted depends on the moisture content and temperature. Periglacial landscapes are formed in those parts of the world which generally remain below freezing. The active layer is the soil near the surface which thaws in the summer. The soil below that, the permafrost is 'permanently' frozen possibly to a depth of many hundreds of metres. The slow down-slope flow of soil and boulders as the active layer freezes and thaws is called **solifluction**. There is usually an absence of stabilizing factors such as plant roots so the solifluction produces angles as low as three degrees. This process forms flat tundra landscapes such as those found covering much of Siberia and Alaska.

Other types of slow flow include dry movement of coarse material an example of which is talus creep, the down hill movement of scree.

Rainwash and raindrop impact will also move particles down-slope. The burrowing of animals and the actions of farmers also assist soil creep. In summary any soil which is disturbed is unlikely to return to an up-slope position: it may return to

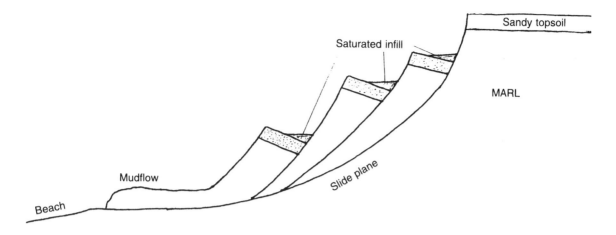

Figure 2.30 Coastal landslip

the same place, but is more likely to return down-slope.

Rapid flow often depends on the amount of water in the moving mass. Saturated earthflows or mudflows have the broadest velocity ranges, from 10 mm per day to 1 m per second.

Further to the classification is rockfall. This is primarily the movement of individual blocks which may momentarily exhibit slide and will then fall and/or bounce. Occasionally major flows take place at angles of greater than 40° on steep mountain slopes or cliff faces. The rock face is usually of sufficient strength to support cliffs of greater height, but the faults and fissures allow blocks to detach.

Sliding occurs when the drift material all moves with the same velocity across a plane, as shown in Figure 2.28. The material may be relatively dry but the plane is lubricated by rainwater. These movements are often slow, fed by erosion at the top and depleted by wind or waves at the bottom. When rapid slides occur they are usually spectacular. One of the difficulties facing civil engineers considering coastal defences is landslips (Figure 2.30). They usually consist of the either near lateral, or rotational movement of blocks against a plane lubricated by water trapped by previous slides.

The debris avalanche is a combination of slide and flow. It starts as a slide over a well-lubricated surface. As the velocity of the slide increases the material breaks up and the parts and any other disturbed material move as rapid flow.

Case Study – Aberfan

At 9.30 am on 21 October 1966 in the South Wales mining village of Aberfan near Merthyr Tydfil a generation of schoolchildren perished in the worst disaster of its kind to ever occur in Britain. As the children and teachers prepared for their day an avalanche of 2 million tonnes of mine waste, boulders, rocks and sludge slid down from a fifty year old coal tip and engulfed Pantglass Infant and Junior school, a few hundred yards from the base of the tip. A row of terraces houses and a farmhouse in the path of the avalanche were also demolished.

The bodies of teachers and children were discovered in the playground and classrooms buried under 15 m of saturated slag and mud. The body of the assistant head, D. Benyon, was found with his arms clasped around five children. They were all dead. The bodies of two girls were found in the playground still holding hands. By nightfall seventy-one bodies had been discovered. The final death toll was 147.

A statement issued by the National Coal Board intimated that the disaster was the result of two days of heavy rain and a spring at the base of the tip which was unknown to them at the time.

Under normal circumstances the tip would have been stable, but no account was made of the effect saturation has on the angle of repose. Unfortunately the person responsible for the inspection and therefore the stability of the tip was a mechanical engineer, and not, as he should have been, a civil engineer. There are a number of ways the collapse could have been avoided.

1 The first and most obvious solution would have been to site the slag heap elsewhere. Unfortunately the topography of South Wales is dominated by hills and valleys so stability would always have been a problem. The other alternative would have been to remove the slag to a safer area. The coal board would, no doubt, have been reticent when faced with the extra cost.

2 When instability is created by excessive rainfall it is wise to cover unstable land with waterproof material either temporarily or permanently. As the saturated soil dries so the angle of repose increases.

3 Ensure that the water can drain out quickly. This is best achieve by incorporating land drains into the slag heaps as they are built up. Any attempt at excavation into a saturated slope is likely to make matters worse. The importance of drainage was stressed as far back as 1927 when the NCB was warned by Professor George Knox, '. . . if you do not pay for the drainage, you will have to pay for the landslip . . .'

2.10 WEATHERING

Igneous and metamorphic rocks are formed under high pressure and at high temperature, and usually in the absence of water or free oxygen. When they appear at the surface they undergo a release in pressure, a variation in temperature and a exposure to oxidation processes.

Weathering is the natural breakdown, or disintegration and decomposition of rock *in situ*.

Mechanical weathering is the breaking down of rock into smaller parts. Initially this leads to fragments of the same, and subsequently to the formation of sediments, e.g. sands.

Frost shattering occurs in rocks which hold water in the pores, or in joints or other crevices. This is also seen in building bricks with high permeability. The surface material absorbs water which subsequently expands on freezing and causes the surface layer to break away.

In the natural environment rock outcrops will often be jointed, and water will enter and often remain as narrow deep puddles. The **freeze–thaw** process widens the gap until a block is separated from the main body of rock. Where the slope is shallow this process will result in the production of **blockfields**. Frost shattering will not occur where the climate is:

● permanently frozen
● very dry
● very wet (plant cover).

Thermal expansion is more prevalent in environments where the diurnal temperature range is extreme, such as a desert. The outer layer of a rock will expand and contract producing strain between the outer and inner sections causing the top layer to peel off, or **exfoliate**. Granular rocks will produce finer disintegration than homogeneous rocks since the minerals expand and contract at different rates.

Chemical weathering is the decomposition which follows chemical change. This takes place in most climates, but is more rapid in a temperate environment in the presence of moisture and vegetation. Hydrolysis occurs in reactions between water and minerals and leads to the decomposition of rocks. Feldspar readily attracts water. The hydrogen ions from the water replace the cations from the mineral (cation exchange) and form clays.

Carbonation occurs where rainwater with carbon dioxide in solution produces carbonic acid, H_2CO_3, a weak acid which reacts with limestones. The carbonic acid reacts with calcium carbonate producing calcium bicarbonate, which is highly soluble.

Chemical weathering is more common:

● in zones of alternate wetting and drying
● where minerals readily react with water.

Biological weathering is the result of the activities of plants and animals and is usually included with the mechanical and chemical weathering. There are, however, chemical and physical processes that only occur because of the presence of organic life. Lichen extract iron from rocks, rotting plants produce humic acid, and plants and animals generally increase the amount of carbon dioxide in soils and therefore the rate of carbonation. A plant growing in a crevice may, if the root is strong enough, split the rock.

The prime factors in weathering processes are the presence of water, and variations in temperature.

2.11 SOILS

A definition of soil depends on the industry involved. Two extremes are farmers and civil engineers. To the farmer it is the material in which vegetation grows so by definition would include humus and nutrient. Civil engineers are responsible for moving soil from one place to another so their definition would be based on ease of transport. For construction purposes a soil is an unconsolidated deposit consisting primarily of rock fragments of a variety of sizes.

If the particles are very small, for example clay or silt, they will stick together in the presence of moisture and form a cohesive soil. A granular soil lacks the very fine particles and will not form a coherent mass.

2.11.1 Soil profiles

Residual soil profiles are typified by the section shown in Figure 2.31a. The parent rock (which may also be a sedimentary deposit such as Tertiary sand) is often weathered so that it appears to pass vertically up into the soil in diminishing quantities. There is a precipitation layer through which water passes carrying some of the organic material and minerals with it from the leached layer. The organic material in the **topsoil** is from nearby plants and animals, and it is that which must be removed before a construction project begins. Transported soils (Figure 2.31b) are more of a problem in that they are unpredictable. River action moves sand, gravel and silt, glacial deposits may appear over almost any other deposit, as will wind blown sand (Figure 2.31).

There are various types of soil each with its own distinct properties and uses based on the type of parent rock and, to a greater extent, on the the distribution of particle sizes. Primary soil types are classified by their particle size as shown in Table 2.2. Sand and silt are sub-classified to indicate the particle

Table 2.2

Soil type	Particle size/mm
Clay	<0.002
Silt	0.002–0.06
Sand	0.06–2.00
Gravel	2.00–60.00
Cobbles	60.00–200.00
Boulders	>200.00

or grain size within the soil type using the prefix fine, medium or coarse. A coarse sand for instance has particle size between 0.5 mm and 2 mm.

A typical soil consists of a mixture of the types shown and fortunately for the farmer and unfortunately for the construction industry, a proportion of organic material.

2.12 SITE INVESTIGATION

Site investigation is the study of the local environment and ground conditions on and around a specific parcel of land. Its function is to ascertain the suitability of a site for the proposed project, or to determine the most suitable project for a particular site. There may be three components, the walkover survey, the desk study and the ground investigation.

2.12.1 Walkover survey

A walkover survey reports on the appearance and characteristics of the site. A brief history of the landscape and possible constructional uses can be found by a variety of means. For instance many local people can remember what land has been used for in the past. The names of roads, farms and public houses give clues to the landscape. The area to the north of Portsmouth Island called Tipnor (North Tip?) is a few metres above sea level. Much of the land there was reclaimed many years ago following landfill. Rubbish found in the excavations includes clay pipes and clay jam jars with metal sprung tops.

In Milton Keynes an estate is named Kiln Farm after the farm it replaced. The old road leading up to the farm is called Clay Hill. There is no need to look in the county archives to conclude that bricks were once manufactured there. Further effort is required however, to assertain the whereabouts of the holes left by the extraction of clay, especially since they are often re-filled with waste.

While it is not possible to be absolutely certain there are a number of topographical features which allow for educated guesses about the soil. For instance large flat areas close to rivers, or on the sides of valleys are usually river terraces and so tend to cover geological units comprising of sand and gravel.

Soil slope angles of greater than around 30° are potentially unstable. Evidence for movement in the surface layer by flow is provided by trees with bent trunks, leaning posts and walls and bends in otherwise straight hedgerows.

Trees and shrubs extract the minerals they require for growth from the soil as a solution of inorganic

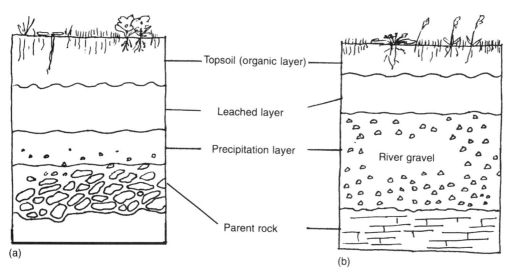

(a)

(b)

Figure 2.31 Simplified soil profiles (a) a residual soil (b) a transported soil

substances in water. The solution is taken into the plant by the roots, up through the stem and branches, and with the necessary minerals removed is passed out through minute cell arrangements in the form of holes in the leaves called stomata. It is also through the stomata that the carbon dioxide is absorbed into the plant, and the oxygen is transferred to the atmosphere. The amount of water removed from the soil obviously depend upon the size of the plant. A large oak may release as much as 650 litres from the soil into the atmosphere each day. When a tree is cut down prior to construction two things happen. The first is that the root space, provided that it has been removed, must be filled and compacted. The second is that the moisture that seeps into the clay soil over the next few years will not be taken up by the tree, but will fill microscopic spaces and cause the clay to swell or heave. This has often resulted in fractured foundations and drains and is a good reason for comparing old maps and photographs with the current situation. The water demand by each species of trees is different. An oak will take up considerably more water than a Stone Pine, so the extent of the soil heave following removal varies with the different species. It is therefore important for the walkover survey report to include the position, height, spread, girth and species of all trees and large shrubs.

If the site has existing buildings they will provide a good indication of suitability of the soil for construction purposes on the part of the site on which they are situated. Brick buildings with cracked walls should be inspected for continuing

movement. Underpinning is indicative of previous foundation failure. Buildings shown on old maps which no longer exist suggest that they became derelict. There are a variety of good reasons for recommending a thorough soil investigation.

Surface water should be mapped so that the position can later be checked with historic sources. A pond or stream which has been filled-in either naturally, with a mixture of humus and surface material, or artificially, usually with either household or construction rubbish, can create serious problems once the project is under way. In either case the soil is unlikely to support a building and consideration should be given to either avoiding that area if possible, or sinking the foundations through the infill to solid soil beneath.

It is not possible to list every detail that should be recorded on a walkover survey. Much of it is down to common sense. At a site on the side of a hill in Dorset it was found that fifty or so years ago it was used as a quarry, so a trial pit and deep bore holes were required. What was not noted though was that the only way onto the site was on foot down steeply paved steps. The result was two-fold. First the machinery to carry out the ground investigation was unable to enter the site, and secondly the construction company needed to buy one of four houses and obtain permission to demolish it to build a service road, an expense the company really should have been made aware of by the walkover survey before they committed themselves to the expense of a full ground investigation.

2.12.2 Desk study

The preliminary information is gained by use of a desk study. As the name implies this does not necessarily demand a visit to the site, but is better carried out in conjunction with a walkover survey. General information required can be found using:

- aerial surveys where appropriate and available maps
- geological surveys and mining records
- adjacent site investigation reports
- position and altitude of adjacent Ordnance Bench Mark (OBM).

The area around the site, as well as the site itself may be the subject of Local Authority restrictions. The site may contain listed buildings or trees with preservation orders. Ancient monuments are perhaps obvious, but to those without an archaeological background, some burial mounds may look like piles of soil which could easily be removed with a mechanical shovel. Part of the site may be a right-of-way or have restrictions imposed because of adjacent properties, possibly interfering with their right to light. The site may in the past have had approval for industrial buildings which, if constructed, would cause problems to the prospective owners of adjacent houses. At many sites access provides no difficulties. At other, more secluded sites it may produce problems of ownership.

Information about the geology of the site may be found in geological maps and memoirs, and soil surveys which are produced by Ordnance Survey and geological institutions such as the Geological Society and the Institute of Geological Sciences. Site investigators must consider local characteristics. These may not be permanent and can vary from flood risk to seismic activity.

Each site will require a supply of gas, electricity, clean water, a drainage system connected to the main sewers, and communications. These together come under the heading of services. Information should be sought from the services authorities and the communications should form an appendix to a desk study. Certain aspects are common to all services suppliers. They are:

1 position/location of existing services
2 depth or height of cables or pipes
3 costs involved in installation
4 location of service authority local office
5 bye-law and other requirements of the authorities.

Gas and electricity are supplied by private companies licensed by the government to supply energy. They will communicate technical information such as the voltage, phases and frequency of the electricity supply, and the calorific value of the gas.

The report should also include waste disposal. This includes the position, depth, bore and type of drains and sewers adjacent to the site and any bye-laws controlling their use. Information regarding the responsibility for the disposal of solid waste, e.g. domestic refuse should also be included.

2.12.3 Ground investigation

All construction projects are supported by the ground. Unfortunately we too often take for granted its stability and resistance to applied loads. As a consequence approximately 50 per cent of all projects over-run and losses of up to 35 per cent of the original tender have been suffered. The primary reasons for these losses are either an insufficiency of ground investigation, or a lack of understanding of the results. Typically on-site problems fall into two categories.

1 Man-made obstructions
  (a) Foundations and services from previous projects on sites where only the superstructure has been demolished.
  (b) Methane pollution and soft ground on reclaimed land.
  (c) Quarry and mine waste which will not support the proposed structure.
  (d) Recent removal of trees.
2 Natural obstructions
  (a) Underground springs which only appear as damp patches or a change in vegetation cover in normal weather conditions.
  (b) High water table on a sloping site.
  (c) Change in the sediments which is not visible at the surface, for example an infilled pond.
  (d) Chemical composition of ground water, e.g. concrete made from ordinary Portland cement is susceptible to chemical attack by sulphur.

That a lack of suitable site investigation exists is indicated by the fact that the National House Building Council, an organization which insures the structure of the majority of new domestic construction, pays out £3–6 million each year in claims related to geotechnical problems.

The cost of site investigation varies according to the project, but is commonly less than 0.5 per cent of the contract price and sometimes as low as 0.1 per cent. However, with a reduction in profit margins to as little as 3 per cent, an additional 0.5 per cent in costs initially appears as a reduction in profit by one sixth. The use of cut price site investigation – or none at all – is a form of gambling. If the gamble pays off the company will have saved a small proportion of the costs. If not the additional on-going costs will inevitably exceed the price of a competent site investigation report. The general view among civil engineers is that costs are generally increased by inadequate or absent site investigation.

The correct foundation dimensions are necessary for two reasons. If the original estimate of the width of a strip foundation was too small the drawings need to be re-done and re-submitted. If the foundation were too big there would be an excessive cost for the concrete.

The extent of the ground investigation will be determined by the results from the desk study and walkover survey, and the type of buildings proposed for the site. Low-rise domestic construction investigation usually consists of a desk study and walkover survey, followed by *in situ* ground investigation using boreholes and/or a trial pit, and laboratory analysis of particle size distribution, plasticity and chemical composition.

2.12.3.1 Methods

The primary function of the ground investigation team is to describe the ground, not to make judgements concerning its suitability for any particular purpose. A trial pit, large enough for an operative to stand in, may be excavated to give easy access to the strata. The use of the hand auger (Figure 2.32) has advantages where the soil is cohesive (but preferably not dense dry clay) in that:

1 it is cheaper than other methods
2 if carried out properly causes minimal disruption.

A series of boreholes quickly provides an overview of the proposed site, and indicates possible further ground investigation requirements.

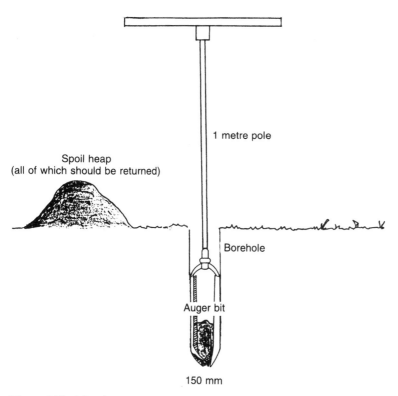

Figure 2.32 A hand auger

2.12.4 Soil description

A site investigation report should use language that can be understood by any construction technician or professional. An architect and a first year assistant site manager may both need to refer to the report. The method used for the field description of soils is therefore standardized and kept as simple as is possible.

The type of deposit is written in capital letters so that it is highlighted on the borehole log. It is usually preceded by the colour which can be arrived at by comparison with colour charts, or by a general impression. The soil on many beaches could therefore be described as 'Orange SAND'. However many soils are a mixture of particle sizes so that a gravel might contain sand, or a sand clay. Sand may be further described by grain size. In the field a small board may be used with particles of known size glued to it for comparison. Otherwise, if the particles are difficult to see they are **fine**, if they are easy to see but cannot be picked out individually by hand they are **medium** and if they are large enough to pick out they are **coarse** grained.

In the field clay may be described by its strength. Table 2.3 below shows the relationship between strength and the application of the field tests.

Table 2.3 Soil strength

Indicative density	Applied field test
Very soft	A ball of clay is placed in the palm of the hand and the fist is closed. The clay squeezes out between the fingers.
Soft	A small ball of clay can be moulded with the fingers into a cube.
Firm	Moulding is difficult and requires a great deal of pressure from the fingers.
Stiff	Clay can only be dented by the fingers.
Very stiff	Very firm finger pressure only produces slight indent.
Hard	Can only be dented by a pencil or nail.

Table 2.4 Soil density

Indicative density	Applied field test
Loose	Easily penetrated by hand pressure on rod.
Firm	Driven in easily by 2.5 kg hammer, each blow driving the rod in greater than 300 mm.
Dense	Each blow of the hammer drives the rod in less than 300 mm and more than 100 mm.
Very dense	Penetrates less than 100 mm with each blow.

Unconsolidated soils can vary from loose, like a dune sand, to dense like a compacted gravel. The field test is simple. A 15 mm steel reinforcing rod is pushed into the borehole to test each deposit. Table 2.4 shows the relationship between indicative density and the appropriate field test.

The method above is obviously subject to variations in the force used by the person with the hammer. A more precise method is to use a penetrometer (Figure 2.33). The mass falling over a fixed distance produces a constant force.

The methods used above are combined to produce a description of soils which make up the strata. This is known as a borehole log (Figure 2.34). A further important piece of information recorded on the log is the altitude above sea-level of the water table.

The depth of each bed and the altitude of the surface of the borehole (Ordnance datum) and all of the strata boundaries are recorded. The number of boreholes to be drilled and the interpretation of the ground between them is a matter of local and geotechnical experience and common sense.

2.12.5 Particle size distribution

Sediments are formed of a variety of particle sizes. A glacial till often contains a broad spectrum. A river sand is unlikely to contain very fine particles since they will stay in suspension until the velocity and density of the water changes at the estuary. In estuarine silts the range is therefore considerably narrower and towards the finer end of the spectrum.

We can also look at the shape of the particles. The grains of sand produced by glaciers are the result of

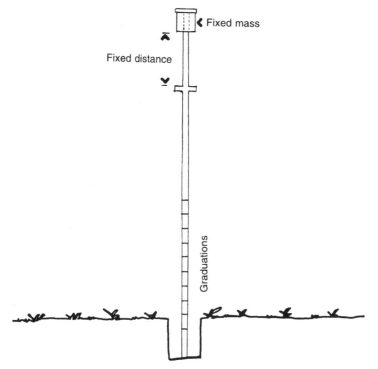

Figure 2.33 The penetrometer

the grinding action and are therefore very angular. A desert sand will be blown across the surface and tend to be more spherical. Larger particles which are moved without rolling, similar to those on a low energy beach, become discoid.

It is therefore possible to get an idea of the depositional environment of a sedimentary deposit by looking at these features.

The variation of grain sizes in a given sample is characterized statistically by separating the sample into classes based on particle diameter using a set of stacking sieves. The frequency of each class is represented as percentage mass and plotted against grain diameter. The simplest method is the histogram which has the advantage of visual representation. However, a smooth cumulative frequency curve is better since it shows continuous variation and allows for statistical analysis. Recording grain size using a linear scale is unsatisfactory since all sands, silts and clays are covered by the first 2mm while pebbles and cobbles range from 4–256mm. Soil engineers overcome this problem by using a set of sieves where the aperture size of each is twice the size of the sieve below and the results are plotted as if they were linear.

Sedimentologists and geologists have a slightly different approach. The latter use a $\log_2$ scale which would form a series such as:

8 mm   4 mm   2 mm   1 mm   0.5 mm   0.25 mm

These values can be represented in $\log_2$ form as:

2^3 mm 2^2 mm 2^1 mm 2^0 mm 2^{-1} mm 2^{-2} mm

so we can use the index to represent the particle size in each case. The only problem remaining is that all sand samples and smaller will have a negative size so we take the negative logarithm in each case.

So the **phi** size which represents 4 mm is −2, 8 mm is −3 and 1/16 mm is 4.

Statistical analysis:

Mode   The most commonly occurring particle size or class (from the histogram).

Median   The grain diameter which corresponds to the 50 per cent mark.

Graphic mean $\mu = \dfrac{\phi 16 + \phi 50 + \phi 84}{3}$

CLIENT	A. CONTRACTOR		SURVEYOR	P. CLARK	
BOREHOLE NUMBER	04		LOCATION	SUPERMARKET KINGS ROAD	MAP REF 705692
METHOD	HAND AUGER		REDUCED LEVEL 13793		DATE 7 JAN 96

H₂0	Reduced level	Thickness	Depth	Legend	Description
		0.5			black organic fine SAND
	13293		0.5		
▼ ⎯ S̄	12593	0.7	1.2	• 1	grey medium SAND with quartz pebbles becoming clayey
S				• 2	dense grey CLAY with angular flints.
	11693	0.9	2.1		
					end of borehole

NOTES

S = saturated
▼ watertable at 0.95 m

samples taken at • 1
and • 2

Figure 2.34 Simplified borehole log

$$Sorting \qquad \sigma = \frac{\phi 84 - \phi 16}{4} + \frac{\phi 95 - \phi 5}{6.6}$$

$$Skewness \qquad sk = \frac{\phi 16 + \phi 84 - 2\,(\phi 50)}{2\,(\phi 84 - \phi 16)}$$

$$+ \frac{\phi 5 + \phi 95 - 2\,(\phi 50)}{2\,(\phi 95 - \phi 5)}$$

Grain shape Although they sound straightforward, roundness and sphericity are easily confused. Both are quantifiable but where the sample is a sand or smaller, measuring each shape would be extremely time-consuming so we estimate the average shape against a standard set of images.

Roundness refers to the smoothness of the surface of the grains. A discus is well rounded as is a cylinder.

Sphericity is a visual determination of the relationship between three dimensions. A table-tennis ball is well rounded and has high sphericity. A discus is well rounded with low sphericity.

2.13 COASTAL PROCESSES AND DEFENCES

2.13.1 Waves

Contrary to appearance a wave is not a small body of water travelling across the surface of the sea, but a small scale rising and falling of the surface. It is, like all other waves, evidence of the energy which has produced it. The formation of waves is directly related to the velocity of the wind across the oceans, and the distance over which the wind and waves travelled (the fetch). The terms used to describe waves (Figure 2.35) are similar to those used in other branches of science.

Three parameters worth noting are:

1 *Period*   time between successive crests (t).
2 *Velocity*   wavelength divided by period (d/t).
3 *Steepness*   waveheight divided by wavelength.

The longest fetches and strongest winds are usually found in the southern hemisphere where wave heights of 20 m are not uncommon. Around the coast of Britain wave heights of 3 m are rare.

2.13.2 Wave energy

Waves may travel for thousands of kilometres before reaching a coastline. As it approaches land the depth of water reduces and at the sea bed the speed of the wave energy is reduced by drag. The energy of the wave is absorbed by the cliff or beach.

In the case of the beach it is the sand or shingle that absorbs the energy or rather converts it into sound and heat. A high energy wave is a good example of the physical mechanisms involved in that exchange of energy. As the wave breaks, usually with an audible crash, there is a rapid forward movement of water – the swash. Sediment from the beach is thrown up into the body of the water and moved up the beach. As the energy is dissipated a short silence is followed by the sound of water moving back under the influence of gravity – the backwash. On shingle beaches this is associated with a loud rattling sound. This has a spreading effect on the beach sediments. If there is sufficient sand large breakers, or high energy waves will produce wide flat beaches, the sort surfers prefer. As the average energy of the waves is reduced so is the swash. Less sediment is therefore pushed up the beach, (Figure 2.36) the backswash is also reduced but becomes the most significant mover of sediment.

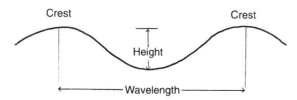

Figure 2.35 Wave terminology

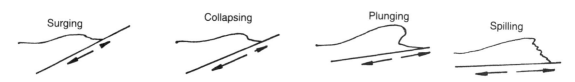

Energy increase with decrease in gradient
length of arrow is proportional to swash and backwash

Figure 2.36 Beach slope and wave type

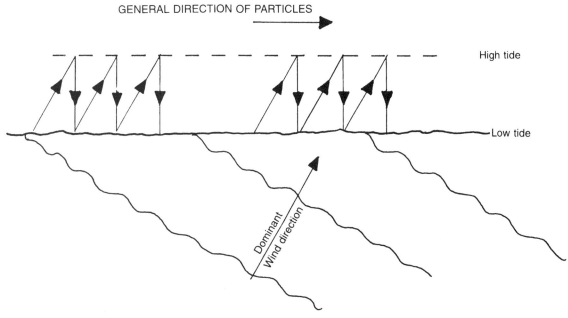

Figure 2.37 Longshore drift

It therefore follows that given a sufficient supply of sediment flat beaches absorb high energy waves and steep beaches are associated with predominantly low energy waves. Not all flat beaches are the product of high energy waves, but it is essential for the coastal engineer to understand the relationship between wave energy and beach dimensions.

2.13.2 Longshore drift

A secondary feature of wave action occurs where the wave direction is oblique to the beach. The particles are forced up the beach by the swash and so follows wave direction. The backwash however is determined by gravity and therefore takes the steepest gradient back to the sea (Figure 2.37). Consequently beach sediments move in a shuffling, zigzag fashion along the beach and, since few beaches are exactly at a right angle to the wave action, longshore drift is common around the coast of Britain.

The direction and extent of longshore drift may change several times during a day, week or month, especially during storms, but does maintain a predominant direction.

If, as has been described above, the function of a beach from an engineering viewpoint is to absorb energy, then the retention of the beach is essential as a protection of the cliffs or dunes.

The sand and shingle of beaches which are susceptible to longshore drift is often retained by placing groynes at intervals along the shore. Once the groyne is in place the effect of longshore drift becomes more obvious as the sediment accumulates on one side (Figure 2.38). The direction of longshore drift is therefore towards the side with the greatest accumulation of sand.

Groynes are a local solution. If a stretch of coastline which was previously unprotected has groynes constructed along it the sand or gravel will accumulate. However, prior to the construction of the groynes the sediment would have continued along the coast. The consequence may be a depletion of beach sediment at a neighbouring beach and the resultant exposure of the land to wave energy.

2.13.3 Beach profile

Beach profile will be determined by a large number of factors which include wave energy, sediment supply, rate of longshore drift and wind direction. A typical beach profile will include many of the features (Figure 2.39) though perhaps not all of them.

Figure 2.39 shows how the larger particles are found in the backshore, often forming a storm beach. Towards the shoreline the beach flattens and is capable of absorbing greater wave energy.

Figure 2.38 Sediment accumulation at a groyne

2.13.4 Coastal erosion

The erosion which occurs around the British Isles is, in the medium term, responsible for loss of expensive land and properties. In the longer term much of coastal Britain will either be defended by large-scale civil engineering sea defences or will disappear.

The primary factors that influence the rate of coastal erosion are the resistance of local geological strata to the predominant wave action and the movement of groundwater. We are fortunate in that most of the west coast of Britain is of rocks which are resistant to wave action while the south and east coasts are generally of softer deposits. Pictures of great waves crashing against towering granitic cliffs of Scotland, Wales and the West Country reassure us of the stability of our island. A secondary, though in many cases equally important, absorber of wave power is the beach. The Atlantic waves which are the source of entertainment for surfers in Cornwall, crash onto the beach lifting particles of sand, pushing them up the beach as previously described. Where igneous cliffs and wide beaches dominate the coastline the need for coastal defence schemes is negligable. The most rapid erosion is found along the south and east coasts where the geological formations are sedimentary or of less resistant rocks. There is so much

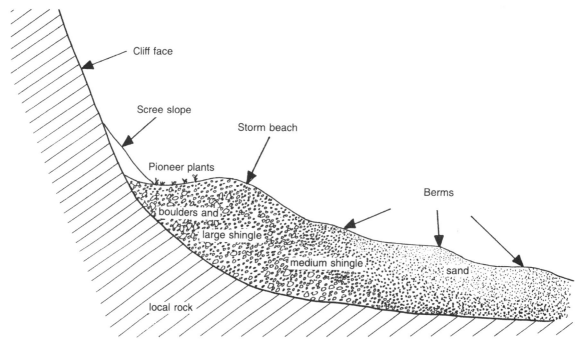

Figure 2.39 Typical beach profile

erosion in some areas that it is statistically impossible to keep up with it.

On the east coast the most spectacular coastal erosion occurs when cliffs and any properties on them crash into the sea. Almost any section of beach in Norfolk or Lincolnshire will exhibit sections of brick wall, parts of drainage systems or lumps of tarmac or concrete. The west Dorset and east Devon coast is more susceptible to landslips which are followed by the slow erosion of the material by water from the land and the sea.

The primary concern must be the difference between price paid to the civil engineers for carrying out the work and the value of the land defended. The coastal land of west Hampshire and east Dorset is almost entirely urbanized so that the cost of the land is high. This makes the cost of defences worthwhile, especially since the primary problem is created by the movement of groundwater. The south coast of the Isle of Wight on the other hand is exposed to high winds and waves and forms cliffs of considerable height with little access to the beach. The land at cliff tops is of a lower value so defence would be considered too expensive. Whether future generations will agree remains to be seen.

As well as being financially acceptable, coastal defences should also be environmentally sound. A good understanding of coastal geomorphology is requisite. Many millions of pounds have been wasted on coastal defences which have been washed away in the very storms they were designed to protect against. Building groynes below the eroding cliffs is relatively cheap and, where possible, very effective. It is extremely important here to take an overview. If longshore drift is interrupted a reduction in the supply of sand to some existing beach, either temporarily or permanently, may be disastrous.

Important though it is to keep waves away from unstable cliffs there is little we can do about exceptionally high tides except to make allowance for them in our designs of sea defences. However, high tides are not only responsible for the rapid destruction of landscape. They are also responsible for the flooding of low lying coastal areas.

2.13.5 Flooding

Another, equally important aspect of coastal defence is therefore to reduce the incidence of flooding. There are a number of parameters which should be considered in the design of flood defences, some of which are more obvious than others:

- low lying areas are more susceptible to flooding
- onshore winds cause an accumulation of sea water at the shore
- low air pressure allows the sea to rise slightly
- different areas of the coastline will have different tidal ranges, e.g. high tide in South Wales is higher than in Cardigan Bay
- coastal flooding is more likely when rivers are in flood.

Case study – North Sea storm surge (1953)

At the end of January in 1953 a low pressure system deviated from the normal path which would have taken it to the north of Scotland and Scandinavia and followed the path shown in Figure 2.40. to an unusual position on 31 January and 1 February 1953. Because of the unusually low air pressure, only 976 millibars, the sea was 0.5 m higher.

The northerly gale force winds over the longest fetch possible in that area created waves 6 m high and caused a further rise in sea level in the southern North Sea. The rainfall produced as the frontal system crossed the country had flooded the rivers and the high spring tide effectively stopped the river water flowing into the sea. The result was an additional 2.5 m of water, on which rode 6 m waves. In southeast England 264 people were drowned. In the the Netherlands 1800 died.

There is firm geological evidence to show that the land around and beneath the North Sea is sinking very slowly, perhaps at about 1 or 2 mm each year. As each decade passes the danger increases. If the conditions of January 1953 were to occur again there was a real chance that parts of London would be flooded. To counteract the forces of nature the Thames Barrier, a complex and expensive flood gate, was constructed.

There is little doubt that relative sea levels are rising. Since the turn of the century it is estimated that a rise of 150 mm has occurred and a further 200 mm is expected by the year 2030. This may not seem much but the subsidence of the North Sea will also contribute another 40 mm giving a relative rise in southeast England of nearly 400 mm. Quantifying the effects of such a relative rise in sea level is almost impossible. Some problem areas are:

1 Sedimentary cliffs which are protected by narrow beaches may at present be pounded by waves once

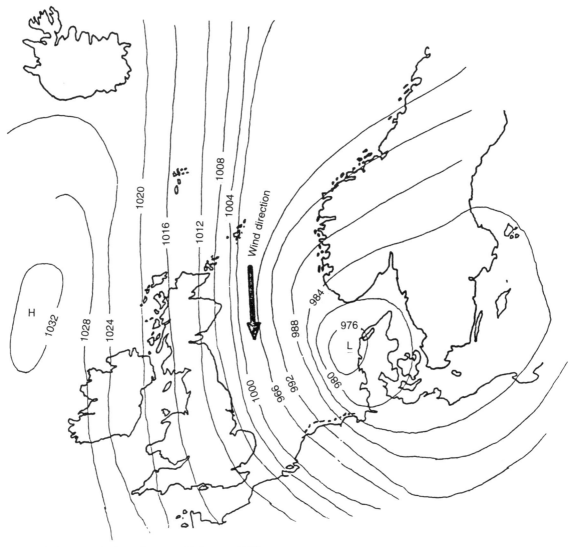

Figure 2.40 Simplified weather map 31 January 1953

or twice each month. A rise of 0.4 m would result in an increase in the frequency.

2 The material between the bottom of cliff and the sea acts as protection against wave action. A rise in sea level could produce erosion of that material.

3 Much of the land close to sea level is on sedimentary strata. If nothing is done to prevent it large areas of land may be eroded. For instance, the Wash will almost certainly increase in size and become less efficient at absorbing the energy of the North Sea storms. East Anglia is predominantly on sedimentary deposits so considerable erosion will take place. The great fear must be that once it starts it will be all but impossible to stop. We have heard of 6 m waves driven by a gale force northerly. Their effect on a cliff face of unconsolidated sediment would be considerable and the resultant loss of towns and communications disastrous.

4 Coastal habitats would be under pressure. Sand dunes and salt marshes will migrate gradually inland as the sea level changes. Those ecosystems unable to migrate because of limiting factors such as roads and railways will simply drown.

The precise cause of the rising sea level is debatable. Many scientists and technologists believe it is associated with the greenhouse effect, warming atmosphere and melting ice. Others believe it is associated with the demise of the mini-ice-age. Others believe it is simply a function of the position and tilt of the earth as it moves around the Sun and that we are simply experiencing a warm decade or two before we head inexorably for the next glacial stage. It is therefore tempting to do nothing rather than spend billions of pounds on sea defences, but doing nothing is such a risk that it cannot be an option. If, by the year 2010 the sea level is still rising, and the warm winters are still with us, the construction industry may quite suddenly find itself inundated with work as panic sets in.

2.13.6 The future

In the very short-term the Meteorological Office has a Storm Tide Warning Service which gives early warning of storm surges in the North Sea. But what of the future? Were our understanding of the natural environment sufficient to allow us to make accurate predictions then our direction would be surer. Unfortunately we do not know whether the atmosphere is going to get warmer or cooler over the next one hundred years. Others factors likely to influence the sea level in the short term include higher winds and deeper depressions, both of which are the possible results of an increase in atmospheric energy. The evidence for a more active atmosphere is as firm as that for rising sea levels. Is it reasonable to wait another few years and see what happens? The only sensible answer seems to be to plan now for the worst scenario. The Department of the Environment suggests that the long-term projections for sea levels as a result of global warming, and the increase in storm surges such as that experienced in 1953, would increase both the area at risk and the frequency of flooding. It is worrying, though not statistically useful to note that in 1991 a flood along the Firth of Clyde, Scotland was described as the highest sea level for one hundred years. An almost identical event occurred again in 1994.

If the worst scenario is accepted we must decide what the possible options are and which of those options is most sensible. Each coastal region must be assessed separately since the geology, weather, rivers and landscape are different, but the national overview must be the main consideration.

1 Do nothing
  (a) In some regions the land may be of little value so the collapse of the cliffs would increase the movement of sediment (by longshore drift). The recipient of the sediment may enhance the retention by increasing the size of structures such as groynes. This would increase the width of the beach and therefore its ability to absorb wave energy.
  (b) Plan for the medium term by compensating land and property owners for their losses. In some areas the cost of coastal defence will be considerably greater than the cost of compensation. In the long term of course this may be regretted since applicants for compensation will continue as long as the cliffs erode.
  (c) Not all movement of sediment along coasts is erosional. Some regions have gained land, Dungeness for example has added around $200 \, km^2$ in 2000 years. It is possible to increase the rate of coastal deposition by environmental engineering, by working with the natural processes and thus increasing the rate of deposition. The common methods involve planting suitable vegetation in the path of sediments, maram grass naturally traps wind-blown sand to form dunes, and salt marsh plants trap fine river sediment before it enters the sea.
2 Construct concrete or natural rock structures offshore to absorb some of the wave energy. This has the advantage of maintaining its usefulness even when covered by water since it would still cause a wave to break and give up much of its energy.
3 Reinforce the cliffs. This is expensive and has a survival time of only tens of years. If it fails when sea levels are higher erosion will be extremely rapid and the land it was designed to protect will be lost.
4 Drain the land close to the cliffs. In many cases the cliffs collapse because of the water they contain. But for the constant removal by the sea of the scree, they would form sloping landscape. If the land were drained the cliffs would be more stable. Whether this is viable depends upon the strata and what other action is also taken, for instance adding extra beach material, man-made or natural, could protect the base of the cliff from the wave action.
5 Construction of continuous sea defences.
  (a) Reinforced concrete walls cost around £2.2 million per km (1995) but if constructed with a good understanding of their effect on the local physical environment they will last around one hundred years.
      Costly mistakes have been made however. In many of these cases the rise and fall of beach material has not been allowed for, neither has the effect of making the beach narrower. In both cases the ability of the beach to absorb wave energy was reduced and the wall undermined.

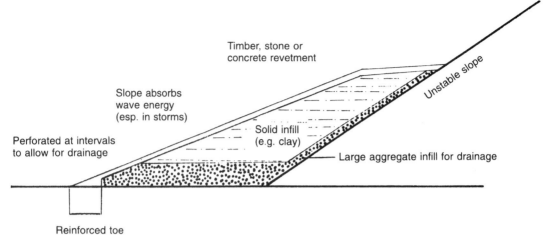

Figure 2.41 Typical revetment

(b) Revetments, (Figure 2.41) are considerably cheaper at around £300 000 per km. and last around thirty years.

This suggests that revetments, at around £10 000 per km per year are cheaper than concrete sea walls at around £22 000 per km per year, but these figures are averages and there are many other considerations, for instance revetments restrict access to the beach and detract from its appearance. This would probably deter holidaymakers causing a local loss in revenue. Concrete walls, while they are entirely unnatural are aesthetically more pleasing than revetments, and often allow pedestrians a good view of the beach and sea. They may therefore attract holidaymakers to the area.

Steel pile walls are not generally used in areas of natural beauty since they detract from it.

Cost benefit analysis shows in the majority of cases that there is no profit to be made from coastal defences in the short or medium term.

3

Construction and pollution

The aims of this chapter are to:

1 provide a general introduction to the concept of pollution and the associated terms
2 provide a general definition of a pollution event
3 show that the terms used are applicable to the construction industry
4 show that, in many cases, taking care of the natural environment is cost effective.

3.1 INTRODUCTION – ECONOMICS AND ENVIRONMENTAL PROBLEMS

Environmental issues are driven by public concern. As the concern increases governments tend to act. As a consequence more legislation is appearing to deal with the protection of the natural environment. Many construction personnel are firmly of the opinion that compliance with the legislation, and taking responsibility for limiting damage on a voluntary basis is not cost effective. As we consider the medium- and long-term examples it seems more than a coincidence that protecting the natural environment often saves money. Take for instance the cost of one wasted brick. An extreme environmentalist may complain about the quantity of CO_2 produced in its production. The practical environmentalist on the other hand can point to the likely costs of waste in terms of sprained ankles, opening new landfill sites and the increased erosion of buildings caused by acid rain – a by-product of energy production. The assessment of the cost of pollution is extremely complex. The formulation of workable policies is simpler. We must decide to use less energy, to reduce waste and create less noise. We often hear such phrases as 'this is not our planet we are simply looking after it for our grandchildren'. However, we are not the first polluters. Our grandparents contaminated land by their industrial efforts. Now that potential development land is considerably more scarce we are in the process of clearing up the pollution they left. To do better than they did only requires a reasonable understanding of the environment, the application of common sense, and a will for change.

3.2 DEFINITIONS OF POLLUTION

There have been many definitions of a pollution event, many specific to the role of the individual or industry. Modern definitions include four concepts.

1 The first is that the event is not part of the normal or abnormal working of the natural environment. Sulphur dioxide is present in the atmosphere in minimal concentrations. In developed countries a major source of atmospheric sulphur dioxide is the fossil fuel power station. In the natural world sulphur dioxide is discharged into the atmosphere in large quantities when a volcano erupts. The former is artificial and so constitutes an act of pollution. The latter is completely independent of our enterprise and as such is not a pollution event.
2 The general response to the term 'pollutant' is that it is 'a chemical'. Our planet is made up entirely of chemicals. Pollution is the result of the release of excessive amounts of either substances or energy.

3 Pollutants are occasionally substances which do not occur naturally. However, the majority do and it is the increased quantity or concentration which is a pollution event. Prior to the industrial revolution the concentration of carbon dioxide in the atmosphere was around 280 parts per million by volume. The value is now believed to be approaching 340 parts per million by volume. The concentration is increased by motor vehicles and power stations and creates an imbalance which, according to current theory, causes the temperature of the planet to rise. If the amount of carbon dioxide is artificially reduced to below the natural concentration in the atmosphere, this too would constitute a pollution event since, according to current theory, the temperature of the planet would fall. Had the concentration of CO_2 remained constant there would be no pollution problem. It is clear then that carbon dioxide is not in itself a pollutant. It is the change in concentration that constitutes a pollution event.

4 While public perception of pollution is usually of a human activity destroying part of the natural environment, we must also consider the pollution of the built environment.

In future there will be an increase in the movement of information between construction companies and other organizations. It is therefore in the interest of all parties to use the same terms when discussing pollution. However because of the subjective nature of this topic and the almost endless number of organizations with an interest in the subject, a brief all embracing definition is a practical impossibility. Given the four concepts above a brief definition which is applicable to the construction industry is:

Human interference with the natural or built environments adversely affects living organisms or structures.

3.3 CLASSIFICATION

When considering a pollution event it is useful to be able to classify the pollutant. A combination of several sources gives the following method.

3.3.1 Character or nature

Pollutants are either substances or energy. It is no surprise that all three natural states of matter are all represented by pollutants. Among the **solids** are those from industrial and domestic wastes. The most problematical of the **liquids** is sewage. The **gases** which cause concern when discharged into the atmosphere include sulphur dioxide, carbon dioxide and nitrous oxides. **Energy** is released into the atmosphere in the form of heat and sound waves.

3.3.2 Source

Almost all industries (and indeed ordinary human activity) are sources of pollutants. Power stations and motor vehicles are major producers of atmospheric pollutants in the form of gases and particles. In recent years considerable attention has been paid to the use of lead in petrol so that the government has levied a greater tax on leaded petrol than on unleaded as an incentive to the motorist to save money and reduce the emissions of lead. An increasing proportion of cars are now manufactured with diesel engines because they are thought to be less harmful, although there is the possibility that they produce equivalent quantities of different pollutants.

When a sound causes irritation it becomes noise. This is a good example of the subjective nature of pollution events. A construction worker may keep a radio playing for the entire working day. To him or her it is a form of entertainment and possibly a welcome distraction. To a local resident living close to the site it may represent a noise which becomes more of an annoyance as time passes.

Major pollution sources are mining, the food industry, the metals industry and sewage treatment works. A complete list of sources of pollutants would include practically every human activity.

3.3.3 Sector of environment

Pollutants affect one or more sectors of the natural environment. Fuel fumes affect the atmosphere, leaked diesel damages the rivers and lakes. Compaction and debris damage the soil. Toxic and radioactive wastes have been damaging the oceans for many years.

3.3.4 Recipient

While the source of pollutants is often from a narrowly defined area, e.g. a power station chimney, the recipients are not. Acid rain, a by-product of fossil fuel power stations, is spread over a wide area. Thousands of acres of Scandinavian pine forests have suffered as a result of acid rain which probably originated in Britain and Germany.

The illegal dumping of toxic waste in rivers reduces the oxygen concentration and so river life is damaged. The Chernobyl explosion discharged a large quantity of radioactive dust into the atmosphere which settled on the vegetation of many European countries. Subsequently many grazing animals had to be destroyed.

All plants and animals are susceptible to a greater or lesser extent to various forms of pollutants. Consideration must also be given to the effect of airborne pollutants on buildings. Section 3.6 deals with this subject in detail.

3.3.5 Effect

The effect of the pollutant on the recipient may be limited to annoyance in the case of an irritating noise, stress if the noise is louder, or even partial or total deafness. River pollution often causes the death of plants and animals which are not tolerant of low oxygen concentrations or increased toxicity. The use of dangerous materials may lead to lung cancer or blood disorders.

Case study – London smog

Arguably the greatest pollution disaster Britain has experienced is the London smog of December 1952. The stable high pressure system, which during the winter tends to cause fogs, lasted from early December for almost two weeks. The smoke from the coal fired power stations mixed with the fog to produce a dense London smog. The air mass over the city remained stationary for more than a week. A concentration of 0.2 ppm sulphur dioxide is generally considered a serious danger to people with bronchitis. In December 1952 concentrations of 1.34 ppm were recorded – roughly six and a half times the danger level. The other irritants, which included sulphuric acid, smoke and smuts constituted a danger which was greater than the sum of their parts. Estimates of the number of deaths are based on comparisons between the number of deaths from respiratory illnesses over the period in question and what would normally be expected. The final estimate was that a pollution event which lasted ten days killed 3500–4000 people.

3.4 OTHER CONSIDERATIONS

3.4.1 Pollution conversion

No industrialized society will ever stop polluting, but we can try to limit the damage by disposing of pollutants by different methods. The disposal of paper and timber on site is normally by removal to a landfill where the organic material will eventually produce methane. An alternative is to burn it. The organic content of the waste will then be converted to carbon dioxide and smoke. Thus, either the land or the atmosphere will be polluted. On a larger scale some cities are overcoming the cost of transportation to landfill sites by building incinerators. This eliminates the pollution associated with the transport of the waste but introduces the problem of emissions from the incinerator.

A similar situation occurs in the choice between a diesel and a petrol engine. Diesel engines produce pollutants but they have two advantages. The first is that diesel engines generally achieve greater efficiency than a petrol engine in an equivalent model. The quantity of fuel used is therefore reduced. The second is in the type of pollutant. Petrol engines produce gases and very fine particles which may damage the lungs and get into the blood. The particles from diesel engines are larger and more likely to be apprehended before they reach the lungs. It is probable that a complete change to diesel engines would result in an increase in illnesses caused by smoke and particles. However, the use of the internal combustion engine to describe pollution conversion indicates a further, sometimes extreme action: minimize pollution by using a pollutant free means where possible.

3.4.2 Biological properties

Bioaccumulation may occur when a pollutant is consumed by an organism near the base of the food chain. That plant or animal is often unaffected by the poison. A higher animal would perhaps eat large numbers of the simple organism and the pollutant may by degrees be passed up through the food chain, possibly to humans, who may be seriously damaged by it.

A second important biological feature is biological decomposition, the ability of microscopic organisms to consume pollutants. This is often advantageous, but can cause problems, for instance when algae consume nutrients in water. In the presence of high concentrations of nutrients they become so successful that they dominate still water by forming blooms at the surface.

The blooms cut out sunlight, and limit the amount of oxygen absorbed into the water, thus threatening other forms of aquatic life. The insects and microbes at work in a sewage treatment works on the other hand consume the decomposing material in liquid sewage which helps in its purification.

3.4.3 Scale and term

Paraquat is a dangerous weed killer with short term toxicity. It is effective because it poisons on contact, but as soon as it touches the soil it is neutralized. Minute doses of carbon monoxide on the other hand may have no appreciable effect on a human being but it stays in the blood for a considerable time, each minor addition building up the concentration. This is clearly a long term problem.

The **scale of release** of pollutants influences the final outcome. According to the definition of a pollution event, the destruction caused by the Dam Busters during World War II was an act of deliberate pollution. The dams were destroyed by bombs, but the buildings were destroyed by the water released by the destruction of the dams. Under normal circumstances water is not a pollutant. On a hot day in a glass or in a swimming pool it is considered beneficial. However, the sudden release of several million tonnes of water constitues an act of pollution because of the sheer scale of the event.

3.5 POLLUTANT CONTROL

There are three methods available to interrupt the passage of a pollutant from its source to the recipient:

1 Perhaps the most obvious and sensible is to take action at the source, but this is not always possible. Suppose a road gang are breaking concrete. The majority of the noise comes from the hammer inside the drill and the contact between the chisel and the road. The drill can be insulated with padding to reduce the intensity of the sound but the contact with the road remains. To limit the sound at the source is possible, but to completely eliminate it is not.
2 The passage of noise from source to recipient could be interrupted by the erection of a soundproof room around the gang but this would be prohibitively expensive.
3 The third option is to supply the recipient with ear protectors, but this would also be difficult in a busy street.

3.6 THE EFFECTS OF POLLUTION ON BUILDINGS

Weathering happens naturally. Throughout the life of a building its external envelope is attacked by a variety of naturally occurring erosive agents. If buildings are to be preserved, protection against natural erosion is necessary. In many cases the pollutants which damage buildings already exist in the atmosphere. Chlorides and sulphur oxides are examples of chemically reactive pollutants which also occur naturally (from sea water and volcanoes respectively). In general, protection against pollutants is achieved by increasing the frequency of maintenance. However, some materials, such as limestone, are more susceptible to pollutants than others. Kiln baked clay bricks, for example, seem to require no protection from pollutants at all, provided they are of reasonable quality to start with.

3.6.1 Pollutants

The decay of building materials, especially natural rock such as limestone, is to be expected. Nothing lasts for ever. There is unfortunately much evidence to show that rates of erosion are increasing especially in urban areas. The changes in the rates of erosion are tied to the increase in atmospheric pollution. It is important to consider the time span involved for each type of pollutant as well as the type of pollutant and its effect. For instance, sulphur and soot increased considerably as the Industrial Revolution got under way and since they affect natural stone and metal many old buildings and monuments have shown increased rates of decay for many years. However, the Clean Air Acts (1956 and 1968) have resulted in considerable reductions in smoke from most sources. Concentrations of associated sulphur dioxide have also decreased since the mid 1950s. However, these figures may be misleading, as the measurements are often associated with urban areas where the effects of industrial emissions are diminished because increased stack heights move the pollutants to other areas where where there may be no evidence of reductions in concentrations.

Unfortunately, oxides of nitrogen have increased rapidly since 1945 possibly as a result of increased car ownership, though NO_x concentrations have steadily increased. Motor vehicle manufacturers have responded with many technological improvements which include greater efficiency and reduced emissions.

The use of fossil fuels has also resulted in an increase in concentrations of carbon dioxide from around 290 parts per million in 1870 to around 340 parts per million in 1990. Despite international concern about emissions, there is no evidence yet (1996) to show that a reduction is taking place.

Measurements of chlorides are not complete but since the primary source has been coal fires we can assume that reductions in concentrations have occurred since the mid-1950s. When air comes across oceans it picks up chloride ions from the salt water so that concentrations of chlorides in rainwater tend to be the result of meteorological changes rather than changes in pollutant emissions.

Photochemical smogs are the result of reactions between intense sunlight and the by-products of internal combustion engines. The products include photo oxidants such as nitrous oxides and low level ozone, which are thought to be responsible in part for the deterioration of plastics and paints. The knock-on effect, especially where paint is concerned, is that its protective function is diminished leaving unprotected material susceptible to attack by other corrosive agents. Ozone is therefore a major problem in hot cities such as Los Angeles and Tokyo. Changes in the concentration of **ozone** in Britain are not well documented because:

- background levels change with the weather
- our summer temperatures are not sufficiently high, or of a duration, to produce photochemical smogs.

The main sources of pollution in urban areas are therefore associated with the burning of fossil fuels, either in internal combustion engines or power stations.

3.6.2 Materials

Chemical and physical reactions between pollutants and building materials are complex and variable.

3.6.2.1 Limestone

Limestones are permeable and often porous rocks which are naturally weathered by frost and rainwater. The degree of erosion by the freeze–thaw cycle, given similar weather conditions, is determined by the size and number of pores. Rocks with large numbers of very small pores are more susceptible to frost weathering than the those with fewer larger pores. Salt weathering is the production of soluble salts which form in a reaction with natural carbon dioxide

and water close to or on the surface of the stone. The reaction taking place is:

$$CaCO_3 + CO_2 + H_2O \rightarrow Ca(HCO_3)_2$$

As well as weakening the surface by taking up limestone (calcium carbonate) and turning it to a salt, crystallization also occurs which causes the salt to harden and expand resulting in mechanical weathering. The introduction of sulphur dioxide into the atmosphere will allow the crystalline salt to react to form gypsum ($CaSO_4 . 2H_2O$), which is easily removed by rain, thus exposing the surface to further weathering. It is easy to see, therefore, that increasing the concentration of sulphur in the atmosphere increases the rate of erosion of limestone buildings. Although there is as yet no firm evidence, we may surmise from the above reaction that an increase in carbon dioxide in the presence of abundant water and limestone will also contribute to an increased rate of erosion.

A further problem occurs when dry deposition of pollutants which contain sulphur oxide builds up on the surface of limestone. At a relative humidity of above 80 per cent, some of the sulphur oxide reacts with water to produce sulphuric acid which then reacts with calcium carbonate and water to produce gypsum and carbon dioxide.

$$CaCO_3 + H_2SO_4 + H_2O \rightarrow CaSO_4 . 2H_2O + CO_2$$

A further consideration is the exposure of the stone surface to rain and wind. If dry deposition is not allowed to occur because of wind speeds, then the erosion may be reduced. (Wind speeds are often higher between buildings than in the open.) However, under those circumstances the surface will be regularly washed by rain. Rain-washed limestone can look cleaner and is therefore often deemed less of a problem, but because the elements can easily attack a clean, undefended surface rates of natural mechanical erosion will be higher. The rate of erosion from pollution in rainwater adsorbed onto the limestone increases because it is then attacking a fresh surface.

The rates of erosion of other naturally occurring rocks are less well known although we expect igneous rocks such as granite to be more resistant to the forces of the natural environment than, for instance, a medium grained iron sandstone. However, in some cases this is not true. The mineral constituents of granite are quartz, feldspar and mica. Quartz is almost completely unreactive, but varieties of feldspar and mica will erode at different rates in the presence of water, leaving a sediment of clayey quartz particles.

Cement based products are more sensitive to carbon dioxide. When carbon dioxide reacts with water, carbonic acid is produced. The calcium in cement reacts with the carbonic acid to produce a salt which is easily removed.

Reinforced concrete is affected by the presence of carbon dioxide and sulphur oxides. The concrete protects steel from oxidation (rusting) but the reaction with atmospheric pollutants weakens the concrete and allows access of unwanted elements. If the steel is then attacked by sodium chloride from sea spray or road salt corrosion will be especially rapid since the products of the reaction take up more space than the steel causing the concrete to crack and exposing more steel to the elements.

3.6.2.2 Metals

The corrosion of metals in the atmosphere is a chemical reaction between the metal and the corrosive agent which usually takes place in the presence of water and oxygen. The oxide formed at the surface represents a weakening of the metal but can also act to protect it from further oxidation by excluding free oxygen. For example, oxidation on the surface of aluminium window frames produces aluminium oxide which is insoluble and prevents further corrosion. In some cases however the oxidized layer is soluble and easily removed by rain water allowing a more rapid rate of corrosion. If the rainwater contains reactants such as sulphuric acid, sodium chloride or sulphur dioxide, a solution of corrosive material will form. Many of the oxides formed by corrosion are hygroscopic, that is they absorb water from the air around them, so it is not necessary for the metal to be exposed to rainfall or damp conditions for corrosion to occur, although metals intermittently exposed to rain will usually corrode faster than those protected from it for example those under bridges or canopies.

Many building materials are affected by acid rain. The emissions of oxides of nitrogen and sulphur form dilute acids and react with natural stone and metals or they may adhere at the surface and react directly with those materials. Sulphur in solution will react with cement based products, as will carbonic acid resulting from carbon dioxide emissions. The damage caused by emissions of those oxides can take place over many years.

3.6.3 Protection of materials

In the first instance designers of buildings should be aware of the pollution problems prevalent in the area of the proposed building. Stipulating a permeable limestone in an area of frequent rainfall which has higher than normal concentrations of carbon dioxide and sulphur oxides is likely to result in court action within a few years.

Some materials, for example **clay bricks**, require practically no maintenance whatsoever and equally are almost impervious to common pollutants. At the other end of the scale is steel. Mild steel should not be exposed to the atmosphere under any conditions. Exposure to a polluted atmosphere would increase the rate of erosion. The reduction in cross section of a railway line of 1 mm per year is common while in the first half of the century greater losses were recorded. A well established method used to protect steel is to paint it. The only real problem is the regularity with which this needs to be carried out. Other methods include galvanizing, zinc or chromium coating.

Copper, lead and aluminium and their alloys are used in the construction industry because they are resistant to corrosion in the presence of water. **Copper** sheets, such as those used for roof covering, are expected to last for 200 years. **Lead** was once the commonest metal in the construction of domestic dwellings but has been replaced for reasons of public health by copper and aluminium alloys. The flashings and other external lead work is resistant to most atmospheric conditions. However, it is not wise to design lead containers such as box gutters which are likely to fill with leaves or other organic material and remain damp for long periods of time. When organic material breaks down a weak acid is produced which will corrode the lead.

Aluminium is interesting in that the oxide film which coats it when exposed is as strong as the rest of the metal and less reactive. This process is imitated in the production of aluminium products by the anodizing process to give a strong resistant finish. The oxidized film is only broken down by strong acids which would corrode other metals much more rapidly. Care must be taken to avoid contact with salt, copper or mercury, or materials which contain them, for example timber preservatives.

Timber requires no protection from atmospheric pollution since the problems caused by high relative humidity and insect attack are far greater.

Cement products such as renders and concrete are adversely affected by, and should be protected from, sulphate attack. Exposed areas are often decorated, or covered in an impermeable membrane. Sulphate resistant cement is used to protect the concrete in foundations.

3.6.4 Financial considerations

The financial cost of repair necessary as a result of pollution damage is often impossible to determine precisely. The first consideration is the extent of the damage, often confused because, in many cases, pollution damage augments natural erosion. Further difficulties arise when a protective coat is destroyed by pollution. In the absence of a coat of masonry paint the render coat beneath would then absorb water and suffer chemical and mechanical erosion. Similar problems arise with the decay of painted windows and doors. Failure of a lead flashing will result in internal damage. The main cause of the secondary damage, it may be argued, is from a natural source, as only the protective materials were destroyed by pollution. The solution is obvious. In areas which suffer from atmospheric pollution regular maintenance of buildings is essential.

The reduction in atmospheric sulphur oxides over forty years has resulted in a decrease in some maintenance costs. It is difficult to assess the costs over the period of time since records tend to be limited to local authority and historic buildings.

Another consideration is the desirability of the structure. A low cost terrace house will usually suffer erosion of the mortar joints (the effect of pollutants on the bricks is not usually significant). The wall may be washed down with a proprietary cleaner, the joints raked out and repointed in a few days by one bricklayer. At the other end of the scale the limestone of the Houses of Parliament (The Palace of Westminster), was cleaned renovated and treated with a stabilizing solution from 1981–1993 at a cost of at least £30 million.

The damage done to stone figures is impossible to measure in financial terms. Their intrinsic and cultural value is often in their antiquity so repair or complete replacement also constitutes a loss. The statues shown in Figure 3.1 are one and the same. Figure 3.1(a) was taken in 1908, 200 years after the statue was erected. Figure 3.1(b) shows the effect of sixty-one years of pollution.

3.7 POLLUTION BY THE CONSTRUCTION INDUSTRY

3.7.1 Noise

Noise, or unwanted sound, is an irritant or annoyance. When compared to other pollutants it is often considered less of a health hazard or danger but the effects of noise, especially over the longer term can

(a)

(b)

Figure 3.1 The effect of atmospheric pollution (reproduced by kind permission of The Open University)

cause illness through lack of sleep or deafness. It is also an unusual pollutant in that the extent of the annoyance may depend more on what the receiver is doing, or expects, than the noise intensity.

By definition sounds are a pollutant to some people and not to others. Loud rock music, the sound of a cockerel at 4 am, your neighbour's car starting and jet aircraft are all examples of sounds. Whether they are also noises depends on the personal preferences and tolerance levels of the receiver. Tolerance may also change with locality. A low-flying jet aircraft is perhaps tolerable or even enjoyable at an air show, but an annoyance when the receiver is out for a stroll in the country.

Simply measuring the loudness or persistence of sound does not give an indication of the disturbance or irritation caused by noise. An excellent example of this is noise caused by transport. There is no doubt that, given the proximity of people to traffic, and the persistence of the noise, transport constitutes a major source of noise pollution, and yet complaints about it are less than one would expect. There are two possible reasons for this. The first is that we accept noise if it is where it belongs. Mid-morning traffic in a busy city street produces a variety of sounds which are in the main accepted. Engines are revved almost continually, horns are sounded and tyres squeal at corners. In a suburban cul-de-sac at night the same sound would constitute a noise.

Secondly, complaints to Environmental Health Officers (EHOs) about traffic noise are almost impossible to confirm because there is usually no particular polluter but a continuous stream of them. It is possible to make a complaint about one specific vehicle by recording the registration number and passing it on to the police or the local Environmental Health Officer.

3.7.1.1 Noise from construction sites

Many industries create noise. There are places where we would expect certain sounds, for example airports, industrial estates and football grounds. These sources are permanent. A construction site is not, so it is perceived as bringing noise into an otherwise peaceful area. It is probably as much for this reason than any other that the construction industry is associated more with noise than any other form of pollution.

The most obvious consideration is the noise produced on site. Sound intensity varies considerably from that of a mixer running to compressed air jack hammers used in demolition work. One of the commonest complaints is the result of one of the quietest mechanical plants on a site. During exten-sive dewatering, pumps will be required through the night especially prior to a concrete pour. If the site is close to a residential area where the nights are generally quiet a complaint to the local Environmental Health Office is almost guaranteed. The genera-tion of noise by construction traffic either at the site or en route is likely to attract complaints especially where the background levels are low. It is certainly worth considering at the earliest opportunity the distance between the site entrance and local houses.

The Noise at Work regulations (1989) are designed to reduce the noise levels for the site and factory workers. If the noise emissions are reduced at source then the local community will benefit and complaints will be less common.

3.7.1.2 Noise transmission and buildings

Near airports and busy roads windows are often double glazed. Double glazed windows are usually considered for their thermal efficiency, but where external noise is a problem they also function as a sound insulator. The optimum air gap for sound insulation is around 170 mm. While this may be sufficient in rooms where there is a background or distracting sound, it is often necessary to increase the sound insulation in quieter rooms such as the bedroom or study by insulating the roof.

Insulation against the ingression of road traffic noise is an ongoing problem. As new roads are constructed or old ones upgraded any nearby property is likely to suffer increased noise. Whether they are likely to be funded for sound insulation is often a matter of contention which is settled by the local Environmental Health Officer with responsibility for noise pollution.

Noise travelling between properties is the cause of many complaints. Complaints about domestic noise may be described by placing the sources in one of three approximately equal categories, 'dogs', 'music' and 'other'. 'Other' includes loud domestic appli-ances, shouting and DIY.

Consideration of sound insulation in building design is therefore important to protect the occupant from external noise, but also to minimize the sound which escapes from buildings.

3.7.2 Ground compaction and ground water

Lorries and other vehicles tend to compact the soil they move across. Compacted soil is impermeable so rainfall tends to remain at the surface until it evaporates.

Chapter 5 shows that ground water moves slowly through the soil, especially below the water table. When a large obstruction is placed in its path such as a foundation the water can build up and cause the builder and home owner problems. Many gardens have damp patches on one side of the building, often made obvious by rapid growth of moss, while the other side is predominantly dry. A more extreme case is the high-rise estate constructed on the side of a steep chalk hill (Figure 3.2). After prolonged rainfall the ground water came to the surface.

Figure 3.2 shows that facing the hill were a series of service doors into which the water flowed. The water flowed through the building and out of the balconies at the front. This building had several other problems of an environmental nature. Finally, the local authority decided that continually overcoming the lack of planning and environmental consideration was not viable, so after eight years the multi-million pound building was demolished.

One of the most difficult naturally occurring pollutants to cope with, especially from large build-

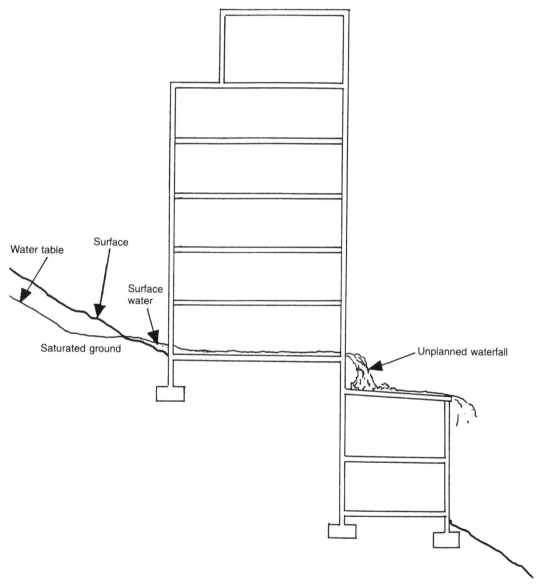

Figure 3.2 Flow of ground water blocked by deep foundations

ing sites and civil engineering projects, is soil. An early process on the majority of sites is the removal of the topsoil. The subsoil is then exposed and eroded by the movement of plant, especially in wet weather. Lorries are capable of collecting large quantities of mud or slurry from a site in the tread of their tyres. Fines and cleaning costs are such that great care must be taken to keep the pollution of local roads to a minimum.

3.8 POLLUTION BY BUILDINGS

The construction of buildings requires an abundance of energy, the majority of which is used in the production of materials. The pollutants produced by the production of energy increase the total energy in the atmosphere, decrease the ozone in the stratosphere, increase acidity of rainwater and increase greenhouse gases such as carbon dioxide.

3.8.1 Heat

Industrial buildings are the primary source of thermal pollution. It is too easy to argue that it is not a result of the design or construction process but of the function of the building. It is, however the role of the designer to consider the function of the building and work with specialists of that and other industries.

A common practice by industries which use water as a coolant is to discharge the heated water into a river, stream or estuary. The resultant increase in temperature of those environments is determined by the quantity released, the amount of mixing and the original temperature difference. Generally, invertebrates are less tolerant to temperature increase than vertebrates. At 20°C most fish are not in danger but invertebrates are. Those animals that feed on invertebrates will have to find another cooler stretch of river where their food supply has not been wiped out.

Thermal pollution in rivers and lakes also interferes with the normal reproductive processes of water life, especially where animals determine the time of year by temperature. Fish will spawn early, and adult insects emerge into an atmosphere which is fatal to them.

Water temperature is also a prime factor in determining the maximum concentration of free oxygen so that animals which can only tolerate well-oxygenated water will move to a cooler region if they can.

If the heat emission is controlled it may have a variety of positive uses. Some edible molluscs, especially oysters and mussels, grow more rapidly and in greater numbers in warmer water. It can also be used to heat greenhouses or pumped into a nearby heating system.

Except in large cities the release of waste heat into the atmosphere is not considered a problem unless it is in association with other airborne pollutants. Whether this will always be the case may be determined by the extent of the greenhouse effect.

Wealth and pollution seem to go hand in hand. The inhabitants of cities which experience anomalously high temperatures consider an air conditioning system essential. The air temperature inside a building is reduced to a comfortable level by a refrigeration unit. This process produces heat which is pumped into the local atmosphere. The hottest weather is usually created by high pressure systems. When the occupants of large cities like New York or Tokyo turn on their air conditioning the waste heat increases the temperature locally. Air in a high pressure system moves slowly down so the rising warm air is trapped and the local temperature increases. To maintain a comfortable internal temperature they turn up the air conditioning!

Thermal inversions occur naturally in the atmosphere. The mechanisms responsible for their existence are varied. Cities in valleys or next to the cool sea are especially susceptible. An artificial inversion (Figure 3.3) may be created when the pollution over a city becomes so thick that sunlight is unable to reach the ground.

The upper section is then heated to a higher temperature than the polluted air below it (Figure 3.3). Photochemical smogs often occur under temperature inversions. Products of fossil fuel combustion, especially petrol, react with sunlight to produce an atmospheric soup of harmful substances.

The first course of action by media and the public following an environmental disaster is to look for a scapegoat. You may remember the structural engineer doing the job of a civil engineer at Aberfan; the designer of Ronan's Point and so on. In nearly all cases it is not the fault of one particular individual. Who, for instance is responsible for photochemical smogs? Is it the responsibility of local industry or the planning authority or is it the fault of the motorist or the petrochemical industry.

3.8.2 Segregation of ecosystems

When a motorway cuts through a previously natural landscape it segregates the ecosystems. A country fox

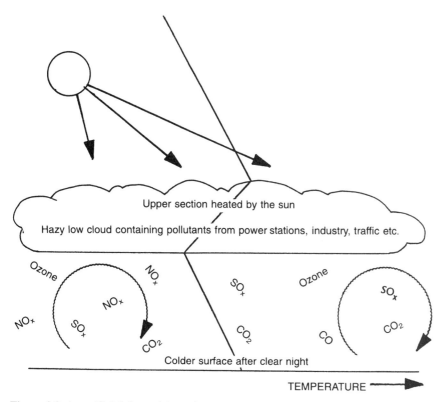

Figure 3.3 An artificial thermal inversion

may survive on a few acres of heathland. If it is cut in half the family of foxes may not be able to survive so their prey population will increase with no control factor until the woodland cannot sustain them.

3.9 POLLUTION LEGISLATION

Until comparatively recently legislation controlling pollution was dealt with as an offence against people and property. If the building company was pile-driving all night the complainant would go to the police and complain that it was a nuisance. Now the police would refer them to the local environmental health office. An Environmental Health Officer (EHO) will visit the site and record the noise level. If the level is above maximum measured limits the company will be warned and if it continues it will be prosecuted.

EC law attempts to impose high standards. The purpose of Article 25 of the Single European Act of 1986 is the preservation and protection of the environment. In 1987 Her Majesty's Inspectorate of

Pollution was established to monitor pollution by radioactive substances, hazardous wastes and the atmosphere by industry. The local authorities' Environmental Health Officers are charged with monitoring and controlling non-hazardous wastes which includes food, industry and domestic wastes.

Since 1987 there has been an increase in EC directives which as a member we are bound to comply with, and as a result our own legislation was in need of updating. The need for new legislation brought about the Environmental Protection Act 1990 (EPA) the main aim of which is to control pollution by prevention.

The primary function of the act is to institute a pollution control strategy which combines the offices of national and local Governments.

The role of the Environment Agency is to monitor discharges into water under its control, which includes rivers, streams and sewers. It carries out this task by preventing unauthorized people from carrying out processes for which they are either not licensed, equipped or qualified. They also issue prohibition notices when there is a risk of serious pollution.

3.9.1 EEC, national and local government

The Control of Pollution Act 1974 and the Environmental Protection Act 1990 give responsibility for the control of pollution to local government and it is through their Environmental Health Offices that complaints are directed, e.g. noise from construction works.

3.9.2 Changes to legislation

Pressure to alter legislation is developing along with governmental concerns about the natural environment. An excellent example is the Code of Practice for waste holders under the Environmental Protection Bill. All industries have to dispose of waste, not least the construction industry.

Statistics about numbers of prosecutions are affected by changing legislation. For instance the number of prosecutions for noise in 1986 was 11 400. During that year the Vehicle Defect Rectification Scheme was introduced which forces the owners of defective vehicles to repair the fault or scrap their vehicles. The number of prosecutions in 1990 was 7900.

3.9.3 Pressure groups

Pressure groups are generally unpaid organizations with a concern for the whole or part of the natural environment. In some European states pressure groups have been so successful that they are represented in national government. In Britain members of the Green Party often stand for election but seldom receive many votes. Probably the most effective of the pressure groups are Greenpeace and Friends of the Earth. These groups are often maligned in the press for the work they do, but many of their policies have become common to other organizations. An example is the acceptance by government that the construction of new trunk roads is a waste of money since they increase road use and create more traffic problems. Governments worldwide have also banned the slaughter of endangered species, which prior to the efforts of those organizations were simply expendable.

Finally, there are a significant number of groups with strong interests in specific plants or animals. They are likely to turn up at motorway construction sites to protest that communities of lizards, snakes, orchids or badgers will be damaged by the work and the product.

3.10 POLLUTED LAND

Polluted land falls into one of two categories, **contaminated** or **derelict**. A precise definition of contaminated land is difficult to find. The general consensus is that the soil on contaminated land contains hazardous wastes in sufficient quantities to pollute (see Section 3.2). The inclusion of the pollution aspect separates contaminated and derelict land. The latter may be defined as land which requires treatment before being returned to general use. It therefore follows from the above that all contaminated land is derelict, but not all derelict land is contaminated. An example of derelict land would therefore be a disused airfield. However, it is likely that fuel was stored on one part of the airfield and that it was often spilled. It is therefore possible for an area of derelict land to contain a smaller area of contaminated land.

3.10.1 Derelict land

Recent estimates indicate that there are approximately 70 000 ha of derelict land in Britain. The problem of ownership, and therefore responsibility, is made all the more difficult by constant changes in ownership of previously nationalized companies and changes to local authorities.

3.10.2 Contaminated land

An increasing proportion of new construction now takes place on land that has been contaminated by previous industrial activity or waste disposal. These activities have left a legacy of aggressive chemicals and gases such as methane which can have serious consequences for materials and buildings and for the safety of occupants.

Building Research Establishment
(December 1994)

Overcoming problems associated with contaminated land is described as either **restorative work** at a site which has been polluted, or **future strategies** and management of existing waste disposal sites.

In Britain there are presently (1995) approximately 40 000 ha of land which has been contaminated by the dumping of dangerous or hazardous wastes. The industrial actions of the past have left in their wake hazardous substances such as mercury, lead, asbestos, arsenic, cyanide and polychlorinated biphenyls

(PCBs). Land at an oil terminal in Aberdeen was contaminated by oil based products in the subsoil, and petroleum based liquids in the ground water. On the site of an old gas works in Bradford the toxic contaminants included coal tar, mercury, cadmium, arsenic and cyanide. The toxicity of the contaminants varies in term and intensity. The cost of decontamination will probably fall between £300–£800 million per annum depending upon how many of the sites are cleaned up each year.

> Containment is currently the most commonly used method for dealing with contaminated sites.
> There are more than 1000 landfill sites in England and Wales where it is suggested that landfill gas control measures are required. BRE is developing guidance on protective measures for new buildings.
> BRE also has sensitive equipment for detecting methane, carbon dioxide and oxygen levels in buildings and on site and for measuring flow rates. Staff have recently been called to measure methane levels in a doctors surgery and CO_2 levels in an old peoples home.
>
> Building Research Establishment
> (December 1994)

Given the nature of some of the contaminants, especially those associated with landfill sites, it is not surprising that fires and explosions are becoming a major problem on reclaimed land.

Case study

In the early years of the twentieth century the private railway companies were building stations and laying down lines to cope with the industrial revolution and the population increase. A station in the South Midlands used a small valley as a land fill site. The paints and varnishes dumped there were covered by other materials for sixty years. The constantly increasing pressure created by the overlying weight of waste caused a spontaneous combustion at the base of the tip. Each year the ground is checked for temperature changes and will remain unusable until something is done to ensure that the fire is extinguished. Without excavating the entire valley this is currently considered impossible.

3.11 DISCUSSION TOPICS

Gravel was extracted from a 1000 ha site until fifty years ago. At that time the local authority issued a licence to a private company to use the site for land fill including hazardous waste. Twenty-five years later a building company bought the land and obtained planning permission for the construction of an estate of semi-detached and detached houses. After the houses were sold the local residents found that a large part of the land, which was missed by the site investigation team, was contaminated by asbestos, arsenic and mercury. Who should be responsible for the cost of dealing with the problem?

3.12 ENERGY CONSIDERATIONS

Energy is used in the manufacture or extraction of all construction materials (Chapter 4). Further energy is required when materials are transferred from place of origin to the builders merchant and to the site, especially where materials are imported. If we are to reduce the environmental costs, and at the same time improve our trading position with the rest of the world, increased use of British products and materials is to be promoted. Availability and quality of raw materials is not a problem. British materials like steel, slate, stone and cement are as good, if not better, than those produced in other countries. The problem has often been price or delivery dates.

Consideration should also be given to energy costs in manufacture. The cost to the purchaser of cavity wall insulation is approximately the same irrespective of the material used. However, the production of polystyrene pellets uses considerably more energy than does mineral wool so that while the householder is saving on fuel and living in a warmer home, any reduction in pollution because of lower fuel use is negated in the early years if polystyrene pellets are used. The implication that buildings which have been designed with a long life use less energy is not necessarily the case. In production timber cladding uses considerably less energy than does steel cladding. It is also a far better insulator. The use of timber also locks up large quantities of carbon which would otherwise appear in the atmosphere as carbon dioxide. It is a matter of balance. Timber products require approximately 20 per cent of the energy of a similar steel component. Consideration must therefore be given to the life expectancy of the components and of the building.

3.13 CONCLUSION

Some of the pollutants from construction processes are by their very nature local and unfortunately rather obvious. Others, like on-site energy consumption, add to the general pollution we create as an industrial nation. Our ability to pollute has not declined. While we have a greater understanding of the long term problems pollution brings, it is to our detriment that the construction industry only responds to environmental problems when ordered to by government legislation.

It is easy to argue that it is not the role of the construction industry to influence the rest of society in the reduction of pollutants. However, planners and designers may move in that direction by:

1 siting shops near people reduces pollution from fuel and reuses land in towns
2 making homes as thermally efficient as is possible reduces the need for power
3 setting examples to employees and site visitors by making clear the intention to reduce fuel consumption on site.

It is surely more than a coincidence that the above items all save the user money.

4

Resources, materials and waste

The aims of this chapter are:

1 to show that, contrary to public opinion, actions taken in response to environmental issues are more likely to be profitable than costly, especially in the longer term
2 to discuss the effect of the collection or abstraction of materials on the natural environment
3 to describe the effect of the use of common construction materials on the natural environment.

4.1 INTRODUCTION

If the construction industry is to continue with its present material use we must plan for a lasting harmony with the natural environment and its resources. If such an arrangement is not possible our natural resources will become increasingly scarce, and inevitably more expensive. For instance:

- a world shortage of copper raises the price of copper tube significantly
- the hardwood component of many joinery products has become more expensive as the forests in which the timber grows are destroyed.

The range and volume of material used in the construction industry is vast. There are the obvious examples, sand, cement, bricks and timber, and the less obvious, steel for the beams and stanchions in steel framed buildings and for the reinforcing bars in concrete. Copper for pipes and brass for fittings. Aluminium and other metals for window frames and waterproofing. Plastics are used for cold water cisterns and tanks, guttering, pipes, for coating cables and for cladding. The availability of raw materials, the impact on the natural environment of their recovery, and the manufacture and transportation which is necessary to convert them into useful components are the subjects of this chapter.

4.2 RESOURCES

All building materials are derived from natural resources. Their collection and reworking to suit our needs often requires a considerable energy input. Timbers for joists and rafters are sawn from logs into uniform shapes at the sawmill and then transported to the timber merchant which is often in a different country. Sawing and transportation consume energy in the form of electricity and fuel. The production of steel beams from iron ore requires considerably more energy than the sawing of timber. In common with other natural resources such as oil, gas and coal, there is a limit to the extent of extraction of raw materials for the construction industry.

4.2.1 Supply and demand

In the first instance there must exist a demand. Without a requirement for concrete millions of tonnes of river gravel does not constitute a resource. It is simply a geological unit. The availability and viability of resources is therefore determined by supply and demand. This much-used phrase requires a detailed assessment of the following conditions.

4.2.2 The technology available

Plant is often designed to extract resources when large quantities are available thus maximizing the speed and profitability of the operation. As the volume of resource at a particular extraction point is reduced the company may elect to close the quarry or mine. The value of the remaining stock increases as the resource becomes less abundant.

4.2.3 The prevailing economic and political conditions

The prevailing economic and political conditions constitute the greatest of the external factors. During the recession of the late 1980s and early 1990s house building was cut so dramatically (Figure 4.1) that both the demand for, and the unit value of, construction resources was reduced.

The implications for any resource extracting company are especially severe since planning is necessarily assessed using contemporary conditions. The value of the resource is obviously an important factor, but where on-going costs are high, turnover (or a lack of it) is equally consequential.

The advances in the technology available for the extraction of metals, especially copper, seem to continue to satisfy demand in spite of fluctuations in price. Deposits which are presently not commercially viable will become viable and the discovery of new deposits will increase, especially in areas of the world which are currently under developed.

4.2.4 Stock and reserves

The stock is the quantity of the resource available. In the case of sand extraction it is the whole quantity of sand under the ground at a particular site. The reserve is the quantity which can be profitably extracted. A cement manufacturer may carry out a ground investigation at a particular site and conclude that there exists a stock of 150 million cubic metres of clay of required quality. However, because of the capability of the machinery and the shape the deposit forms under the ground, extraction of all of the resource is inevitably impossible. The reserve is represented as a proportion of the stock, the magnitude of which depends to a large extent on the resource value. If the resource is very valuable greater effort will be used to extract it than if it is not.

4.2.5 Renewable and non-renewable resources

Non-renewable resources occur where natural replacement is so slow that it is deemed insignificant. We often hear mentioned the difficulties which will beset us when we run out of oil. In reality, of course, we will never 'run out' of oil. As the stock reduces the cost of recovery will increase and be passed on to customers. Would you drive to college if the price of petrol increased ten fold?

The construction industry uses considerable quantities of non-renewable resources, especially sand, gravel, stone, clay and chalk, and since these rocks are readily found near the surface of the earth there would seem to be an infinite quantity available. However, the quarrying of raw materials can only take place on a commercially viable scale. In the long term we may

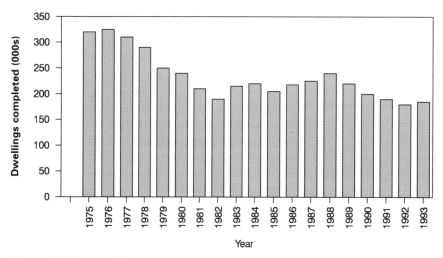

Figure 4.1 House building over eighteen years

regret leaving stocks in the ground which are difficult to abstract. The basic premiss of **supply** and **demand** is that as the stock decreases, so the price increases.

An excellent example of a renewable resource for the construction industry is timber. In the main softwoods grow relatively quickly. They are sufficiently mature to be felled within approximately 100 years, depending on the species. Timber producers can therefore maintain a commercial stock by constant planting. Hardwoods, on the other hand, grow comparatively slowly. Traditionally demand has considerably outstripped **sustainability**. This is especially true of tropical hardwoods and has therefore placed them firmly on the non-renewable list. However, recent attempts to better manage timber production in tropical countries may lead to improved supplies and therefore lower future costs. The International Timber Trades Organisation suggests that tropical hardwoods will be from **sustainable resources** by the year 2000. This may be optimistic on a global scale, but is a step in the right direction.

4.2.6 Conservation

Conservation is an emotive term which means different things to different sections of the community. To the politically minded it has traditionally been associated with resistance to social change. To the resource engineer it applies to the maintenance of scarce resources. More recently the term has become synonymous with the protection of landscape and ecosystems. On closer inspection, however, it becomes apparent that the three interpretations are closely related in spite of the apparent conflict which exists between politicians, the business community and environmentalists.

4.3 MATERIALS

Traditionally, construction materials are from either timber or geological deposits. There is a fundamental contradiction in the gaining of materials, in that society demands on the one hand cheap bricks, aggregates and timber, and on the other deplores the loss of landscape as forests are felled and replanted, and quarrying operators continue with abstraction. It is not an exaggeration to suggest that the first technician involved with construction materials is the geologist. Site investigation for the production of clay bricks for instance will begin where the clay outcrops in abundance. Assessment of the volume and suitability of the material and the operation of the quarry are all part of the geologists remit.

Finally she or he is responsible for restoring the quarry for other use.

4.3.1 Sand and gravel

In the natural environment sand is found as a variety of types which are specified by the mode of transport or deposition.

Wind-blown sand is typified by a desert or dune sand. The energy of the wind is (usually) fairly constant within a certain range so that only particles of below a certain size are lifted into the air. The very smallest of those particles are blown high into the atmosphere while those remaining are transported just above the ground until they settle. Wind-blown sand is therefore a fine sand with little deviation in particle size.

Flowing water is more capable than wind of moving particulate material. As sand in a river valley is transported the finer particles tend to stay in suspension so that any which forms a deposit will be coarser grained than a wind blown sand with fewer particles at the finer end of the spectrum. River sand, therefore tends to vary from medium to coarse grained.

Sand and gravel are usually extracted from either:

- river valleys
- beaches
- glacial deposits (see Section 2.4.3)
- the sea floor.

The majority is from the geological units beneath river terraces.

The environmental impact of extraction comes from two sources. The first is that the scale of the operation tends to leave a hole in the ground which may amount to tens of millions of cubic metres. The second is determined by the location of the site and the roads available. Sand and gravel are intrinsically so cheap that the majority of the costs are derived from the haulage and on-site labour and machinery. It is therefore important for the extracting company to use the largest lorries possible. The use of such vehicles in country lanes creates problems of excessive wear and tear, noise and atmospheric pollution.

We could be forgiven for taking the view that there is an abundance of sand and gravel. Unfortunately, for the deposits to be of any use to us they are required to exist in quantities of greater than 1 million cubic metres. In short there must be sufficient to offset the start up costs, for example the installation of machinery. Since a large part of the cost of sand and gravel is the transport, the distance the materials are transported affects the price. Currently, therefore, shortages tend to be a local problem. It is likely that this will change as sites containing sufficient quantities become scarcer. If the abstraction of aggregate from terraces continues without check the valleys of our major rivers will become more water than land. There seems to be an irony in that as urbanization continues and the land around our rivers is built on (most of our towns and cities are close to rivers) the scarcer, and therefore more expensive, will the earth's resources become. How long will it be before sand and gravel is sufficiently valuable that buildings will be demolished to open quarries?

Beach sand represents a lesser volume but more continuous supply. Because of the movement of sand by wave action a large hole in the beach will soon be refilled. We have already discussed one of the problems resulting from our interference with longshore drift (Chapter 2). Another example seems to have occurred on the Sussex coast. Up until the mid-1960s the beaches from Bognor Regis to Littlehampton were wide and sandy, and attracted tourists in large numbers. Since then sand has been extracted from the beach to the east of Bognor Regis resulting in a depletion as far east as the mouth of the River Arun. The once sandy beaches are now rock platforms. The effect on the leisure industry is impossible to measure accurately.

Case study – Start Bay

Towards the end of the nineteenth century close to a million tonnes of gravel was removed from the beach for use in the construction of the naval dockyard at Plymouth. Unfortunately, Start Bay (Figure 4.2) does not experience significant longshore drift. In terms of sediment movement it is a closed system. Virtually no sediment enters or leaves the bay as a result of natural causes. Following the extraction the level of the beach fell by 4 m exposing the cliff to the erosive action of the waves. By 1957 the coastline had retreated by around 6 m. With no evidence of rejuvenation of the protecting beach many of the residents abandoned Hallsands. Much of it has now either fallen into the sea, or into ruin through neglect.

4.3.2 Cement manufacture

'Cement' has many meanings. The cement referred to here is the powder used in the manufacture of mortar and concrete. There are many varieties of cement each

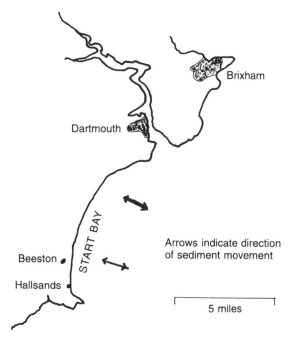

Figure 4.2 Start Bay, South Devon

designed to cope with different tasks. This section will deal with the production of ordinary Portland cement (OPC).

Portland cement was invented by J. Aspdin in 1824 as a replacement for hydraulic lime. One school of thought suggests that he was trying to manufacture a lime mortar the colour of Portland Limestone, and another says that the similarity of colour was simply a coincidence. Prior to the invention of OPC other cements were available but they were expensive and unreliable. It is likely that Aspdin's intention was to produce a more reliable product which could be manufactured at a significantly lower cost than existing products, and in that he was successful.

The basic constituents are lime, silica and alumina oxide. The lime is from limestone and the silica and alumina are found in clay. A common method of cement manufacture (Figure 4.3) is by mixing a clay slurry and ground limestone, at a ratio of around 1:3. The slurry is dried, then burned in a rotating kiln to produce a clinker. The clinker is passed to a mill where it is ground to a fine powder. A small quantity of gypsum is added to prevent it setting too quickly.

The obvious place to site a cement production works is where both clay and chalk outcrop. The limestones are usually either Carboniferous, Jurassic or Cretacous deposits and where they interbed with clayey deposits, such as shale, some of the largest

Production of cement by wet process

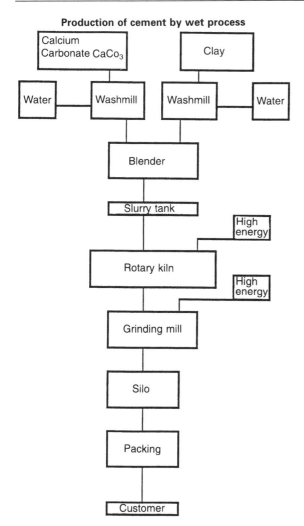

Figure 4.3 Cement manufacture

cement works are found. The most common, accessible and easiest to work sources of limestones however are Cretaceous Chalk. In southern England, therefore, cement is produced where chalk and clay deposits meet, for example the Jurassic clays of south Bedfordshire and the chalk of the Chiltern Hills.

It is an energy intensive process which uses up large amounts of mineral resources. Unfortunately, it is also fundamental to the construction industry. Cement is an excellent example of the material which should not be wasted because:

* waste increases on-site costs
* if less were used there would be a reduction in energy requirement and therefore pollution
* it could reduce pollution at the site of manufacture.

The production of any fine powder is likely to cause health problems, especially respiratory illnesses.

4.3.3 Bricks

In southeast England bricks are manufactured from the Cretaceous Wealden clay. In central England Triassic clays are used. In the West Midlands and northern England most of the bricks are manufactured from Carboniferous clay. The majority of bricks, however, are manufactured from Jurassic clays. Particularly useful is the Oxford clay series which was deposited towards the end of the Jurassic period and now outcrops in a band stretching from Dorset to the Lincolnshire and Yorkshire coasts (Figure 4.4). These deposits were formed around 150 million years ago when the east coast of Britain was an abundance of creeks and marshes. The fine particles which formed the clay were transported by rivers to the coast, where they mixed with small fragments of decaying seaweed and marsh plants, then out into the warm shallow sea where deposition took place. The organic material has, over the millions of years, converted to an oily fuel in the clay.

At the quarry excavators extract up to 25 000 tonnes of clay each week. The processes which follow

Q–Quarternary
Te–Tertiary
Tr–Triassic
Ju–Jurassic
Ca–Carboniferous

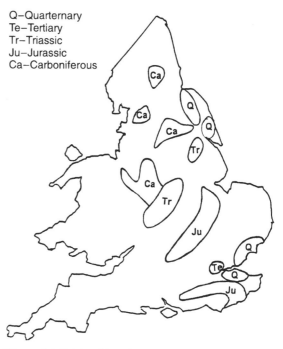

Figure 4.4 Brickmaking clays

depend upon the type of brick. One of the least expensive bricks available is from the London Brick Company (LBC) Fletton or Common. After screening, the brick is shaped by pressure moulding. In the manufacture of the wire cut brick the moisture content should be 22–24 per cent so that it can be passed through an extruding machine. The length of clay is then cut to the appropriate lengths by a series of wires.

Handmade bricks are produced by throwing the clay into a mould. The excess is removed using a wire. Bricks of almost any shape can be produced using this method. The bricks are then dried initially for twenty-four hours in a warm dry place, then in hot air for three days. During the firing stage the organic material ignites and the temperature continues to rise without additional fuel. At 1000°C the rise in temperature is arrested by allowing cold air into the chamber. For the next thirty hours a temperature between 900–1000°C is maintained. This system reduces pollution associated with fossil fuel use, but emits sulphur compounds from the clay into the atmosphere.

There is obviously a high energy requirement in brick manufacture, depending upon the calorific value of the clay. The London Brick Company requires, as a rough guide, a total energy use of 900 MJ per 1000 kg of clay, or 1800 MJ per thousand bricks. This figure is bound to be on the low side of a national average because only lower Oxford clay is self firing.

4.3.4   Blocks

Traditionally the manufacture of many of the building blocks used by the construction industry has been tied to the way fossil fuel power stations operate. Early in the twentieth century workers at coal fired power stations loaded only lumps of coal into the furnaces. The remaining powder and splinters, known as breeze, was used in the production of building blocks. Later some power stations went over to burning coke, the waste product, clinker was also used to manufacture blocks, but because of the colour and texture they were also, though incorrectly, called breeze blocks. Recently the solid fuel power stations changed to a system where the fuel is crushed to a powder to produce a more controlled and efficient burn. The resultant waste, pulverized fuel ash (PFA), is currently the basic component for the majority of modern lightweight blocks. PFA blocks have many advantages over blocks manufactured from other aggregates. The most obvious environmental advantages are:

1 No additional quarrying or mining is required. This cuts down the associated problems as previously outlined and has the added benefit in that every tonne of PFA used in the blocks means a tonne does not have to be disposed of as waste on landfill sites.
2 PFA blocks exhibit considerably greater thermal efficiency than bricks. The U-value of a cavity wall constructed using two skins of brickwork will be around 1.8 W/m²K. If the inner brick were replaced by a PFA block of the same thickness, the U-value will be 0.9 W/m²K. Thermal efficiency in buildings reduces energy cost and demand.
3 Because they are lighter, they are easier to transport and handle on site.

4.3.5   Timber

Until, and to a lesser extent since, the industrial revolution, the most powerful nations all possessed an abundance of timber. The variety of wooden products has been proportional to the variety of species available. Longbows made from yew, warships from oak, cartwheels from elm, cricket bats from willow all helped to shape the character of a nation. Most of Britain's houses were built almost entirely of timber until urbanization and overcrowding resulted in many towns and cities suffering 'great fires'. The industrial revolution did nothing to ease the demand, and it has escalated to such a degree that the current world demand for timber is around 1000 million tonnes per annum.

Technologists tend to classify trees as either hardwoods or softwoods, which are traditionally denoted by the shape of the leaf. In the northern temperate climate hardwoods are broad leafed angiosperms, or flowering plants, which produce a timber with a density of about 700 kg per m³. Softwoods have cones instead of flowers to distribute their pollen, needle shaped leaves and a density of about 400 kg per m³ or less.

Tropical hardwoods are not classified in the same manner. Balsa, for instance is an extremely soft wood with a relative density of about 0.1, yet it is from a broad leafed tree.

The demand for tropical hardwoods, for example teak, sapele, mahogany and afromosia, is greater than our ability, or perhaps desire, to replace stocks. This leads to a reduction in the number of growing trees and an increase in price.

When timber, or any other plant, grows it removes carbon dioxide from the atmosphere. The carbon is used in the photosynthesis of the cells and much of the oxygen is released into the atmosphere hence the reference to tropical forests as the 'lung of the world'. Any reduction in the number of plants capable of converting large amounts of atmospheric carbon dioxide gas into timber will inevitably lead to an increase in atmospheric carbon dioxide, a greenhouse gas.

While felling trees for timber is certainly not the only cause of deforestation it is a contributory factor.

Buyers must also be assured when ordering timber that they are not buying an endangered species. An architect should not specify a tropical hardwood without first checking that it is from a sustainable source. It has been suggested that western governments should ban the importing of all tropical hardwoods. There are two reasons for avoiding this approach:

1 The trees would immediately become close to worthless. This could result in the forest being cleared for agriculture.
2 Tropical hardwoods are found in poor countries. Reducing their exporting potential would only make their people poorer.

Over the last few hundred years there has been an increase in demand for hardwoods, and, in Britain little evidence of attempts to increase the stock. It is now so expensive that it is no longer used as a structural timber and with the exception of a minority of doors, windows and staircases there are few uses for hardwoods in domestic construction.

The structural timber in modern houses is exclusively softwood – pines and firs – which are from renewable resources notably from northern Europe, USA and Canada.

4.3.6 Slate

Slate is a metamorphic rock derived from a predominantly clayey or silty sedimentary rock. In Britain the best examples are found in North Wales (Cambrian and Ordovician), Cornwall (Devonian) and Lake District (Silurian). The pressure and heat associated with metamorphism causes plate like minerals to align at a right angle to the direction of the force. This creates a cleavage plane along which the rock can be split. Slate is unusual because the cleavage plane is more obvious than the bedding plane, though the latter does provide a ripple on the surface which is generally considered an attractive feature.

Slate is also used in the manufacture of simulated roofing slates. It is ground to a powder and mixed with a resin and moulded using a natural slate as a pattern. The product is a realistic copy of the original at a competitive price.

4.4 POLLUTING MATERIALS

4.6.1 Paint

The use of solvent based paints such as undercoat and gloss is becoming a cause for some concern. Other countries are already taking steps to control the sale and therefore production. In Britain, however, we lag behind.

When the solvent escapes from the paintwork into the rooms it becomes a part of the atmosphere we breath. The effects of inhalation of solvents in any amounts are serious. It is believed to result in respiratory and neurological disorders, especially in painters. Unfortunately the public demand a glossed surface on doors and architraves, and solvent based paints provide such a shine.

At this time (1996) there is much activity by paint manufacturers aimed at overcoming the problem before restrictive legislation is introduced. ICI are commited to producing a full range of water based paints by the end of the century. By that time the only remaining solvent based paints will probably be for external use only.

4.4.2 Urea formaldehyde

Exposure to formaldehyde vapour can cause a variety of illnesses including irritation to ears, eyes, nose and throat, severe headaches and depression. Physical contact with formaldehyde is known to cause skin irritation. More recently it is suspected that contact may cause cancer.

The use of urea formaldehyde as a cavity foam insulation should not be considered unless the cavity can be guaranteed to be completely sealed, including at the eaves. Customers who have purchased the product to reduce their fuel bills, and then complained of irritation or illness, have been told to ventilate their homes at all times.

Because of the water resistant nature of the product any material in the cavity such as mortar, or a wall tie sloping inward, will become a bridge for dampness. Incorrectly filled cavities have the same effect. It is also suggested that as it dries, UF foam shrinks producing small canals to aid the passage of moisture.

In the majority of cases insulation of a cavity wall is a sound proposition both environmentally and economically. This is especially true of detached properties where the ratio of wall to window area is at its greatest. There are a number of ways of sensibly achieving satisfactory U-values. The use of urea formaldehyde is not one of them.

4.4.3 Insecticides and fungicides

Wood is a vegetable and all vegetables are food for something. Some wood boring insects attack wood as it grows or shortly after it is felled. Their larvae live on the starch in the sapwood. The larvae of the common furniture beetle (woodworm), death watch beetle and house longhorn beetle, bore their way through seasoned wood leaving a 'sawdust' on the floor and a perforate pattern on the wood.

When wood is attacked by the fungi *Serpula lacrymans* (dry rot) a white slimy cotton wool like substance appears on the surface. When the fungi has taken all it can it disappears leaving a dark dry rotted timber often with deep cuboid cracking.

Coniophora puteana (wet rot), is the most common of the parasitic timber fungi. It is found in damp areas in forests and in damp buildings. The best treatment or preventative measure is to ensure that the moisture content stays below 20 per cent. Normal ventilation should prove adequate.

Wood preservatives are classified as either tar oil, water borne, or organic solvent.

The commonest tar oil is creosote. The only real problem is the smell which is difficult to remove. There is also some concern that burning timber treated with creosote will produce a toxic smoke.

Water borne preservatives are often based on toxic metals such as copper, chromium or arsenic.

Organic solvent preservatives are toxic substances, soluble in white spirit, with a pungent odour which will pollute anything capable of absorbing the vapour including clothes and food. All pesticides and fungicides are a potential danger to the natural environment and their users and should only be used:

- if there is no alternative
- by a competent or, if necessary, a qualified operative
- in compliance with the Health and Safety at Work Act and the COSHH regulations.

4.4.4 Asbestos

The primary use of asbestos in the construction industry is as fire resistant material. It is made up of microscopic fibres which, when inhaled, constitute a serious health risk. It is associated with two debilitating and terminal illnesses, lung cancer and asbestosis. Operatives working with fibrous asbestos should be completely protected and conversant with all current regulations, notably by the Health and Safety Commission and Construction Design and Management. At the British Rail works in Milton Keynes those removing asbestos from railway carriages resemble spacemen rather than railway workers.

4.4.5 CFCs and HCFCs

Chlorofluorocarbons (CFCs) and hydrochlorofluorocarbons (HCFCs) are the refrigerants in most air conditioning systems. CFCs are also used in the majority of foams used for insulation purposes.

They are also believed to be a major cause of ozone depletion. The European Community introduced regulations which banned the production, supply and importation of CFCs at the end of 1994. This is expected to be followed by a similar regulation of the use of HCFCs.

The effectiveness of HCFCs and CFCs is not questioned so their replacement will present construction service design engineers with the challenge of finding a refrigerant which is 'ozone friendly' without posing a new threat to another sector of the environment. Replacements are likely to cost more to produce in the first instant, or will be less efficient as refrigerants. The likely contenders at this time are:

1 Hydrofluorocarbons (HCFs) are similar to CFCs and HCFCs in chemical structure and efficiency, though more expensive to produce. While they are 'ozone friendly' they remain potentially environmentally unsafe in that they are powerful greenhouse gases.
2 Hydrocarbons can be used with safety in small air conditioning units and domestic refrigerators and freezers. They have low pollution potential compared with CFCs and HCFCs. They are also easy to come by in that they are produced by the petrochemical industry. The drawback is that they are flammable and therefore not suitable for use in large scale air conditioning systems.
3 Ammonia is neither a greenhouse gas or ozone depleting but it can be used as an effective refrigerant. Unfortunately, it is both toxic and flammable so that the design of the refrigerator units in air conditioning systems will have to be upgraded.
4 Carbon dioxide was commonly used as a refrigerant before the introduction of CFCs. The reason for its replacement was its lower efficiency. If more energy is required to run air-conditioning units then more fuel would be used and more pollution produced.

In the long term the use of air-conditioning units will depend on whether technology will be available to produce a pollutant free system.

4.5 WASTE

Estimates of the waste produced by the construction industry are currently around 20–35 million tonnes each year depending on the economic climate and political whim. The cost to the industry is indicated by the accepted suggestion that 10 per cent of all materials delivered to sites leave it as waste.

Reducing on-site waste has an obvious financial benefit to the company and several environmental benefits.

1 The production and transportation of materials creates global and local pollution in the use energy.
2 Disposal of waste costs are in transportation and tipping. In some areas, especially large cities, the distance covered by lorries dumping at distant landfill sites is around 200 miles.
3 Approximately 50 per cent of landfilled material is from the construction industry. A reduction in on-site waste would therefore reduce demand for new landfill sites.

On-site waste is either through accidental breakages or predictable waste. The former may include a damaged bath, a crushed window frame or a brick with a chipped face. All accidents are avoidable and most are costly. Accidents on construction sites which involve materials also produce waste which has to be disposed of. The cost of disposal is increasing year by year so that there may come a time when the accidental destruction of a low cost material would be more of a burden because of the disposal cost that the replacement cost.

Predictable wastes including timber off-cuts, unused mortar and concrete, damaged bricks and blocks and broken roof tiles. It is part of a quantity surveyors job to include such items in the price, and to include the cost of clearing the site and disposing of wasted materials.

Some predictable wastes can be reduced at the planning stage. Planners and designers should always be on the lookout for opportunities to reduce costs and lessen the environmental impact at the same time. There are many examples and each project is different. One example is to ensure whenever possible that cut and fill when used are of equal volume, or the excess material is of value elsewhere on the site.

When roads are resurfaced the unwanted asphalt road planings have to be disposed of. Around 8 million tonnes of planings are disposed of annually. The lack of suitable disposal sites in large cities could therefore lead to excessive transport costs. It may become economically viable for local authorities to construct recycling centres thus reducing the volume of transported waste.

Often, bricks rejected because of their appearance are suitable for use in foundations. Waste materials such as bricks and blocks can also be used as hard-fill. Too often bricks are considered expendable, especially in small numbers. This is probably because they are ordered by the thousand. However, a good facing brick is currently around £450 per thousand, i.e. 45p each. Three bricks would equate to a pint of bitter and one costs about the same as two tins of baked beans. Another consideration is the firing costs of new bricks. London Brick Company are currently expending about 900 MJ on each tonne of bricks at the firing stage. The mass of a clay brick is 2 kg, the energy used is therefore approximately 1.8 MJ per brick.

If the company has only one employee on the site and the rest are sub-contractors, the collection of the bricks and tidying of the site is often neglected. It is clear that a labourer responsible for such a task could save the company more than he earned.

There are a wide range of materials used on construction site many of which constitute hazardous wastes. A random sample may include paint, cement and plaster, steel bands used for packing, plastic sheeting and reinforcing bars.

A high proportion of accidents on sites are caused by operatives treading on, or falling over, waste materials, or coming into contact with hazardous wastes. This can result in higher insurance premiums for the company and a loss of earnings for the employee. If the employer is deemed responsible for the proximity of the waste to the employee, as may be the case where hazardous wastes are concerned, the employer is likely be held liable for the employee's loss of income.

Demolition materials are often recycled or reused, especially those from domestic dwellings constructed before the 1930s since they possess a higher value because of their age. Some materials, for example timber, cast iron baths, bricks and slates, may be reused in their original function. Some, such as bricks and concrete may be recycled as hard-fill. Others, such as rotten timber, broken glass and damaged joinery products have no real value.

4.5.1 Recycling

Recycling is the removal of material from waste so that it may be reprocessed. An example is the collection of paper and board which is recycled to produce a slightly inferior paper suitable for newspapers and toilet rolls.

The recycling and reuse of otherwise waste material has many benefits which include the conservation of valuable resources and therefore the protection of the natural environment. The cost of recycling is often assessed in terms of energy. When a resource is plentiful, and therefore inexpensive, the cost of recycling is often financially unattractive. However, as that resource becomes scarce and the price rises, the recycling costs becomes more attractive.

The increase in the use of concrete since the 1940s means that the production of waste concrete is increasing. The aggregates used in the manufacture of concrete are traditionally gravel and sand, and crushed rock. In some areas gravel is becoming expensive because of the lack of local stocks and increases in transportation costs. More buyers are now taking an interest in the technology which exists to crush unwanted concrete so that it may be separated into particle sizes and reused.

The retrieval of useful material on site also reduces the volume of material to be transferred to a landfill site. In some large cities this alone makes the proposition viable.

4.5.2 Reuse

Reuse is the extraction of materials so that it can be used again for the purpose it was originally intended. This is especially prevalent where older materials are believed to be structurally better, or to have extra value because of their antiquity. The Victorian cast iron bath, for example, costs considerably more second hand than it did new. A good example of reuse is the sale of second-hand building materials especially to DIY enthusiasts. Second-hand bricks are usually about a quarter the price of new.

The reuse of timber by second-hand sales is useful in that it reduces the demand on the natural environment. When the tree grew it locked up the carbon from the CO_2 in the atmosphere. When the tree rots or is burned the carbon dioxide returns to the atmosphere. On the down side, the nails in the timber usually have to be removed and the timber treated with a preservative which increases the real cost. Where labour is cheap, or apparently free in the case of DIY projects, the use of second-hand timber saves money and increases resources and protects the atmosphere.

Around 10 million tonnes of concrete and masonry are crushed and reused each year primarily as hardfill. Again haulage is a key factor in the cost of hardfill so the site of use is often close to the site of demolition, or where the cost of disposal is high thus holding down the price. Around 80 per cent of asphalt planings are reused as fill, but as yet only an insignificant amount is recycled.

Metals are always candidates for recycling or reuse, either by sale as second-hand materials, in the case of beams, or by melting down to make new products. Aluminium has a high energy cost at the production stage, i.e. when the bauxite is refined. Initially it is relatively expensive. However it has a low melting point which makes a useful candidate for recycling.

Ferrous metals have been reclaimed for many years and account for the greatest mass of recycled metal. Almost half of the iron and steel used has been reclaimed. The greatest sources are the two million unwanted cars which are scrapped each year either through old age or accident, and the six million domestic appliances such as washing machines, refrigerators, freezers and cookers.

The economic and environmental viability of reuse and recycling are impossible to separate. Attempts are often made to belittle the processes by suggesting that the financial cost of recycling a component is greater than the financial cost of a replacement. The environmental cost of disposal should also be considered. There are also public costs which are almost impossible to assess, for instance dealing with pollution and opening new landfill sites.

Unfortunately, we cannot stipulate that all materials can be recycled or reused. Some, because of physical or chemical properties are simply unsuitable. They may be contaminated by oil or concrete, the removal of which would constitute an unacceptable additional cost. Others, facing bricks for instance, are often required in large quantities. This can influence the price and therefore the viability of reuse.

4.6 HOLES IN THE GROUND

Open pit extraction is by far the most common form of rock and mineral extraction for the construction industry. The impact on the landscape may be obvious long after the quarrying process is finished. If the water table is close to the surface there is little alternative but to allow it to form an artificial lake. If the hole is dry it will be filled usually by domestic waste.

4.6.1 Landfill

The extraction of the earth's resources from quarries creates holes in the ground which can be measured in millions of cubic metres. They are used extensively and landfill sites receive the majority of the refuse in Britain. The disposal of domestic and industrial waste is a serious problem, especially in large cities.

The use of landfill as a means of disposing of refuse causes concern, especially to the people who live near the sites. The operating problems include keeping the rubbish on site, especially in strong winds, and the smell which sometimes occurs, especially in wet weather. A long-term pollution problem is the mixing of ground water with the waste. This produces a polluting liquid called a leachate which must not be allowed to enter the adjacent water table. The floor and walls of the site must be made impermeable. One possible danger then is that the leachate will overflow at the surface and pollute many acres of soil and vegetation. This can be overcome by introducing a drainage system which takes the leachate, as it approaches the surface, to some other means of disposal. If the toxicity of the leachate is acceptable it may be disposed of in a foul drainage system, otherwise it will need to be contained and treated, or permanently containerized.

A further problem associated with landfill sites is the production of methane. The Ibstock Brick Company has made use of the methane at two of their plants by using it to assist with the firing of their bricks.

We have learned from experience that landfill site management requires a good understanding of the natural environment and civil engineering.

Once satisfied that there are no remaining environmental problems the site may be grassed and returned to agriculture or the local community.

Figure 4.5 A reclaimed gravel pit

4.6.2 Leisure

The introduction of several acres of water into an area where little existed previously provides excellent leisure facilities. The quarries to the north of Ringwood, Hampshire (Figure 4.5) are filled with water and are now the home of a sailing club, a watersport club, and several species of wildlife which would not otherwise have been attracted to the area

4.7 CONCLUSION

In many ways we deliberately pollute the surface of the planet we live on by our normal everyday actions. We will not stop polluting, but we can try to limit the damage. The disposal of household waste is an example of a situation where we have choices and with experience and intelligence we can decide on the most appropriate method of disposal. We cannot unfortunately simply choose the method which is least damaging to the environment. It would be useful to be able to put all radioactive waste in a space craft and fire it at the sun, but we must consider cost and practicality.

5

The artificial water cycle

The aims of this chapter are to:

1 show that the efficiency of the artificial water cycle is entirely dependent on the condition of the natural water cycle
2 show that water supply and sewage treatment are major considerations for planners
3 introduce the function of the water industry and its place in the built environment
4 outline the fundamental concepts of hydrology and hydrogeology.

5.1 INTRODUCTION

This chapter is of an introductory nature. The general or ideal conditions will be discussed. Examples of the relationship between the water cycles have shown on several occasions how a lack of understanding of the natural environment costs construction companies large sums of money.

5.2 WATER CYCLES

The hydrological cycle (Figure 5.1) is an imperative mechanism for almost all life on earth and for many of the physical processes. That terrestrial plants need a constant supply of water is unquestionable but they are generally served better by water that soaks into the soil and moves slowly through it rather than rainfall which saturates the soil or flows across the surface.

Weathering is, to a greater extent, a function of available moving water. The minerals are washed from the rocks and may go into solution or be transported as smaller particles to the sea. The erosion of igneous and metamorphic rocks, and the subsequent formation of sandstones and clays is a direct result of the hydrological cycle.

It is not suprising therefore that each branch of science describes the cycle in a slightly different way. Geologists tend to think of it in terms of a weathering agent, while to biologists it supplies plants and animals with moisture and carries the appropriate nutrients. To the construction industry it is a natural process which must be taken into consideration especially where large-scale projects are in the planning stage.

Figure 5.1 shows the way in which water is circulated from oceans, land based water bodies and plants into the atmosphere. Atmospheric motion then transports the moist air. The proportion which forms clouds over land stands a good chance of forming precipitation as rain, snow, mist or fog.

At the surface there are four modes of transport available to the water.

1 Overland flow occurs when the ground is not capable of absorbing water, generally either because it is impermeable, or because the rainfall is so heavy that it is saturated. In Britain it is not common for significant quantities of water to move in this way. It is generally restricted to trickles.
2 The second in order of importance to the construction industry is evapotranspiration. Evapotranspiration is the vertical movement of moisture as a result of either evaporation from surfaces, or the result of the biological transpiration of water from the roots, through the plant and into the atmosphere via tiny pores in the leaves.
3 The third mode of transport occurs as the water is soaked into the ground. It passes through the soil

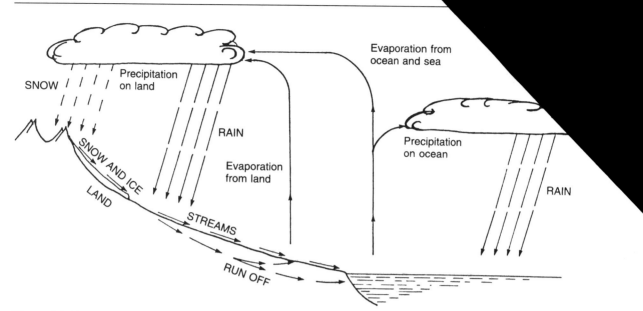

Figure 5.1 The hydrological cycle

and upper layer of rock or sediment until it reaches the water table. Below the water table the soil is saturated by ground water.

4 When groundwater reaches the surface it produces lakes, ponds and streams. They feed the rivers which return the water to the sea.

5.3 GROUND WATER

Approximately 75 per cent of the earth's fresh water is in the form of ice, at the poles and on mountains. Of the rest 0.3 per cent is in lakes, 0.03 per cent is in rivers and the same amount is atmospheric moisture. The remainder is in the ground.

The precipitation that falls onto the land which is not lost through evapotranspiration will seep through the upper layers of the soil until it reaches an area where the soil is saturated. The upper limit is the **water table**, and the water below it is **ground water** (Figure 5.2) The water table is adjacent to the surface relief occasionally coming closer, or even coinciding with the surface. This occurs in a variety of places, notably behind sand dunes, on flat topped hills and in river valleys where areas of marshy land are often visible.

The ground beneath our feet can be classified in terms of water content. The nearest layer is seldom saturated but water must percolate through it to reach the water table. The next layer is intermittently saturated. The water table rises in response to a period of prolonged and heavy rainfall and falls following a drought. Reports that the water table is lower than it has been for many years usually coincide with dry river valleys and hosepipe bans. If the water table continues to fall rationing may be introduced. As the dry spell continues the flow of the river will decrease until, in small streams, the loss through evaporation becomes a significant factor. At that stage the water table has fallen to below the bed of the river. The lowest band is that of permanent saturation, the base of which may be hundreds of metres below the surface. In the bottom of valleys the permanent saturation band reaches the surface and rivers and lakes occur. Although it is described as permanently saturated the boundary between intermittent and permanent may be moved artificially, for example, when excessive water is extracted from boreholes the boundaries will fall locally.

Ground water moves through the permeable rocks and soils under the influence of gravity, but because of the microscopic pore space in the deposits velocities in the permanently saturated zone are generally low. One tenth of a millimetre per hour is common. The majority of water entering a river is from groundwater. The fact that many rivers continue to flow during a drought is testament to both the low velocities, and the huge quantities of groundwater. A river with a cross sectional area of 20 m and a velocity of $0.5\,\mathrm{m\,s^{-1}}$ has a discharge $10\,\mathrm{m^3s^{-1}}$ which is close to

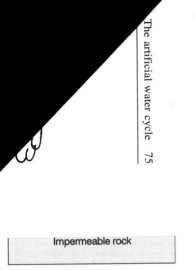

Impermeable rock

Figure 5.2 Ground water

a million tonnes of water per day, all of which is from the ground unless the river is fed by a lake or snow capped mountains.

Question 5.1

If water flows horizontally through chalk at $0.1\,\text{m hour}^{-1}$, How long will it take to flow the 40 km from the catchment area to Piccadilly?

Answer

45.6 years

5.3.1 Permeability and porosity

It is important to differentiate between porosity and permeability. Permeability is the capacity to allow fluid to pass through material, in this case a rock or soil. A coarse gravel such as is used for the production of concrete would be described as having high permeability. Clay on the other hand is almost impermeable. Rocks such as granite are not porous but do contain cracks and joints which allow the passage of water.

Porosity is the percentage of voids in a material. In loose sand it is around 40 per cent. In conglomorates the gaps between the particles are filled by smaller particles so that the porosity drops to less than 20 per cent. Generally the greater the distribution of particle sizes, the lower is the porosity. Porosity is a measure of the quantity of a material which is made up of pores. Some porous materials such as sheet poly-styrene used for insulation of cavities are very porous, but totally impermeable.

Question 5.2

A fresh water aquifer is 6 km square and 500 m deep. The porosity is 30 per cent. Approx-imately how many tonnes of water does it contain?

Answer

1.8×10^{10} tonnes

5.3.2 Aquifers, springs and wells

Springs occur when a layer of permeable rock overlies an impermeable rock. The most typical example is shown in Figure 5.3. Here the rain falls around Upton house and enters the soil. It seeps out where the bedding plane between the sand and clay outcrops.

When the permeable soil or rock overlies an impermeable rock which is either in a syncline, or impounded in some other way, the saturated area is an **aquifer**. The aquifer could be fed constantly (Figure 5.4), or it could have been deposited and then

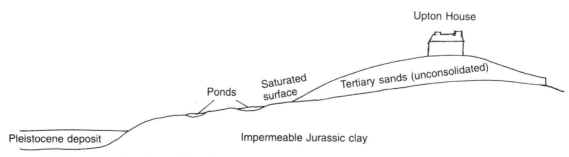

Figure 5.3 Formation of a spring – Upton House

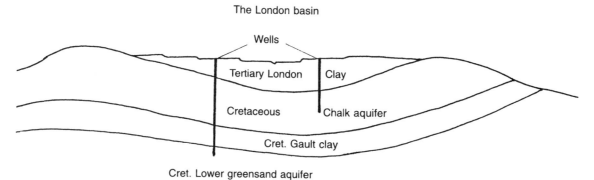

Figure 5.4 Artezian well from chalk aquifer

covered. There are aquifers containing water that is several millions of years old.

Wells are simply holes in the ground that extend either into the intermittent or permanent saturation zone. If it is only to the intermittent zone there is always a danger that during a drought the well will dry up.

An **Artezian well** occurs when the aquifer is sandwiched between two impermeable layers and the catchment area is above the well top. This is the case in the Thames valley. The syncline is bordered to the north by the Hertfordshire Downs and to the south by the North Downs and their western counterparts in Surrey. Following the formation of the syncline the Tertiary London clay was deposited forming an impermeable cover, while the chalk downs are still exposed to the elements. When the fountain in Piccadilly was first constructed it was driven by hydrostatic pressure which is a function of the height difference between the chalk downlands and the surface of the well. Now more than 1000 wells draw water from London's aquifer.

5.4 WATER SOURCES

The water we use comes from one of two sources. The first and most obvious is surface water such as reservoirs, lakes and rivers. The second is water held in aquifers. Nationally the management of water resources is dependent on a detailed understanding of the potential yield in times of drought and flood. The availability of groundwater on the British mainland is dependent upon regional geology. Britain can be separated into two distinct geological areas. The 'hard-rock' area is typified by the abundance of igneous and metamorphic rocks and the associated hills and mountains. The 'soft-rock' areas are typified by the gentle hills associated with central and southern England. Sandstones and limestones make the best aquifers so much of England has a good supply of groundwater. The exceptions are Cumbria and Northumberland in the north, and West Devon and Cornwall in the west. However, underground water is also found in the cracks in igneous rocks so lesser quantities are found in the 'hard-rock areas'. However, the topography and high rainfall of Wales, Scotland, and the north and west of England ensures an adequate supply of surface water.

5.4.1 Rivers

In many areas, especially those previously described as soft-rock regions, there is a real concern that the quantities of water abstracted from rivers will cause them to periodically dry up. This concern is especially relevant where sewage treatment works are using rivers and streams to transport and dilute effluent (see Section 5.11)

5.4.2 Boreholes

Modern boreholes have evolved from the hand dug wells of the last century and before. The steam engined pumps have been superseded by computer controlled electric motors which, as far as is possible maintain a supply sufficient to meet demand. Abstraction rates of 30 million litres per day are achieved at some works.

The speed with which water can pass through an aquifer is principally determined by the permeability of the water-bearing material. A loose sediment of large particle size is more permeable than a sandstone or chalk so water will pass through more quickly. Because of the generally low ground water velocities abstraction at a point tends to reduce the level of the water table locally (Figure 5.5). If the rate of

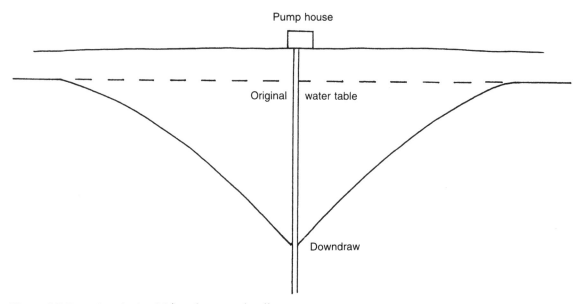

Figure 5.5 Downdraw in the vicinity of a pumped well

abstraction is increased the initial response is for the point of the cone to lower. The width of the cone will then increase at a rate dependent on the properties of the strata and the quantity abstracted. It is common for the drawdown to be measured at observation wells some distance from the pumped well.

Abstraction over many years would, provided there was a sufficient resource and movement of ground water, continue without difficulty. The problems occur when:

- abstraction is increased causing a lowering of the water table
- the ground water is not recharged by (usually winter) rainfall.

In both cases the altitude of the water table can be significantly reduced.

Overdevelopment in terms of water supply is the result of too many people needing too much. In London the number of wells has increased with demand so that abstaction has for many years exceeded replacement by natural replenishment. Since 1820 the water table has fallen by almost 100 m. As well as drastically reducing stock this causes an increase in energy costs because the water is pumped up. It may also increase the dissolved mineral content since mineral rich water is nearer the bottom of an aquifer.

Contamination of ground water may occur, especially if the water table is close to the surface and the ground is contaminated.

Ground water contamination can also occur near the sea (Figure 5.6) where an aquifer contains fresh and salt water. The fresh water is less dense so it floats on the salt water. If it is drawn off faster than it can be replaced the sea water will take its place. A lowering of the water table near the coast could therefore lead to a **saline intrusion**.

Something similar has happened in many estuaries. At low tide the exposed sedimentary deposits once supported springs. When the water table fell the action was reversed so that at above low tide sea water flows into the water table contaminating many of the wells in that area.

This problem has caused the abandonment of wells close to the estuaries of the Mersey, the Thames and many smaller rivers.

River discharge can be affected by reductions in the water table. Section 5.3 showed how ground water is the primary source for rivers and keeps them flowing even during long dry spells. The water table adjacent to a river is often close to the top of the bank (Figure 5.7) so that a drop in the water table equivalent to the depth of the river will cause it to dry up.

5.4.3 Reservoirs and dams

The social requirement for an inexhaustible supply of water is paramount, especially during dry periods. Since much of our ground and river water resources are used at capacity there is an almost constant need

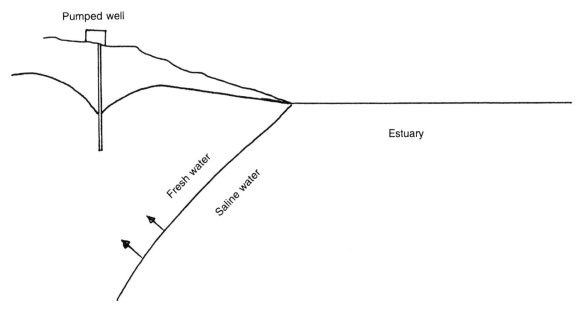

Figure 5.6 Ground water at the coast

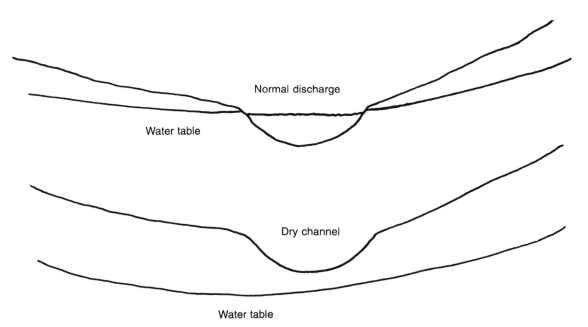

Figure 5.7 The effect of the water table on fluvial discharge

for the construction of new reservoirs or the enlargement of the existing. The function of the vast majority of reservoirs in the UK is to impound the flow of feeding streams. They are therefore called impounding reservoirs. There are thousands of reservoirs around Britain, each one responsible for the loss of part of the original natural environment and its replacement by another. Many are constructed in areas of natural beauty and on National Trust land, sometimes destroying important habitats. It is argued that the introduction of an artificial lake into an area is a benefit to the local environment and is an acceptable trade-off for the drowning of a valley, although this is only sustainable where the reservoir is small since most of the gain is at the lake side. It is doubtful that such an argument can be sustained in the case of the 7.5 mile long Kielder Reservoir in Northumberland.

Once the need for an increase in water supply is determined potential sites will be considered. The suitability of the sites will be compared by consideration of the following features.

1 Water quantity. Estimation of the quantity of water available is necessarily determined using the worst scenario. Calculations of river discharge may be based on the figures for the driest three year period, the worst drought in fifty years or in exceptional cases the worst in one hundred years. Once the design team is satisfied that the supply of water to the reservoir will not run out during dry years, they must consider the storage capacity of the proposed sites. This is initially calculated using Ordnance Survey maps and field investigation of the topography.

2 Quality of gathering ground. The presence of minerals or organic material in the catchment area could be undesirable or harmful. For instance large quantities of peat will colour the water brown. Although there is nothing chemically or bio-

logically wrong with peat stained water it is considered unacceptable by the public.

3 Impermeability of reservoir. Presence of permeable strata at the dam site or in the area of the reservoir may cause leakage. It is practically impossible to assess the volume that will be lost so permeable strata should be covered with clay or another fine material which would in time seal the pores. In exceptional cases concrete is used.

4 Communications. In all cases there will have to be roads sufficient to carry the construction team and often large quantities of construction materials, especially concrete and steel reinforcing bars. This will also act as a service road once the reservior is finished. If the water is for direct supply then pipes to the water treatment works, or the area to be supplied, will be required.

5 Altitude. It is advantagous if the reservoir is at sufficient altitude to allow gravity to move the water from the reservoir to its destination. If it is not it may be pumped to a secondary reservoir at a greater altitude but this obviously costs money and increases the use of power. Most modern power stations do not shut down at night so it is sensible to use the power then as demand is lower than supply and electrical energy is cheaper.

6 Depth of water. Water in reservoirs should be, as far as is possible, free from organic material. The

Figure 5.8 (a) Butress dam

Figure 5.8 (b) gravity dam

Figure 5.8 (c) curved gravity dam

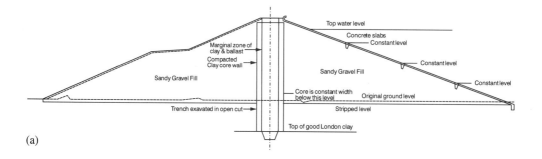

(a)

(b)

Figure 5.9 (a) Typical cross section of earth dam; (b) earth dam at Grafham Water

upper few metres, the photic zone, receives light and can therefore support microscopic plant and animal life. The bottom of the reservoir, the benthic zone, is where the remains of most of the dead plants and animals will finally settle. This constant source of food is consumed by animals of the benthic community. In shallow lakes the photic and benthic zones are close together (in ponds they coincide). Reservoirs should therefore be as deep as the local landscape will allow.

7 Suitability of geology for dam foundation. As more and more reservoirs are constructed the best sites are being used up so that the constant increase in demand and the associated problems with abstrac-

tion from pumped wells and rivers suggests that sites which have been rejected in the past will become viable.

5.4.3.1 Types of dam

There are basically two types of dam.

- **Concrete dams** are, in effect, walls constructed in the path of the stream. The force exerted by the impounded water is acting horizontally so they are strengthened by including either buttresses (Figure 5.8a), by arching the dam, or designing so that no part of the dam is in tension (Figure 5.8b). The latter is a gravity dam. It is also possible to include aspects of each design in the same dam, for example an 'arched' gravity dam such as the Baitings Dam constructed by Wakefield and District Water Board (Figure 5.8c).

- **Earth dams** consist of a central core which is embanked on both sides (Figure 5.9) at a slope which will remain stable when the reservoir is filled. The material for the embankment is generally local, although clay seems to be preferred. The core of the dam is usually of a firm impervious clay, but if that is not available locally concrete may be used.

5.5 WATER QUALITY

The Environment Agency has responsibility for, among other things, the quality of water in our rivers. They carry out monitoring of the water regularly. They are also interested in the quantity of water since it dilutes effluent discharged from sewage treatment works. The waterways in England and Wales are classified into the following water quality groups.

- **Good quality, Class 1a** Water of the highest quality suitable for extraction.
- **Good quality, Class 1b** Lower quality than above but same uses.
- **Fair quality, Class 2** This requires more advanced treatment if used to produce potable water.
- **Poor quality, Class 3** Moderately polluted water which is only of real use for low grade industrial purposes, i.e. as coolant.
- **Bad quality, Class 4** Grossly polluted.

The 1960s and 1970s showed a steady improvement in river water quality. Unfortunately, since then the number of kilometres classified as Class 1 has fallen from 69 per cent to 64 per cent, while Class 4 is unchanged at 2 per cent.

The main reasons for this apparent deterioration is the hot dry summers. If rainfall is reduced the water quality will reduce also. It is also probable that, as in many other cases where environmental monitoring is applied, as our concerns increase, so does the effectiveness of the monitoring. If survey methods become more stringent as they have in some regions, an apparent downgrading will occur.

Ground water is only monitored in connection with its use as potable water. The main long term concern currently is the use of agricultural fertilizers such as nitrates and phosphates. Nitrates are very soluble and move through the upper soils and into the aquifers comparatively rapidly. Phosphates on the other hand react with minerals and organic material in the soil and are intercepted before they reach the water table. Phosphate levels may become serious from other sources such as washing powder.

5.6 SCALE OF ABSTRACTION

The quantity of water removed from our rivers (output) is determined in many cases by the amount of rain that falls in that particular river valley (input). It follows therefore that Wales and Scotland have greater resources than southeast England but the latter has the greater need.

Thames Water, for instance, supplies London with 1.3×10^9 l of potable water each day. For this quantity the dimensions of the main have to be substantial. The recently constructed London 'ring main' has a diameter of 2.54 m. In many cases the water is collected from the river valley and transported to high ground using electric pumps. Large quantities require proportional amounts of electricity to pump the water around. This may represent a substantial part of the ongoing costs as well as an additional source of atmospheric pollution so modern methods reduce the use of electrical energy as far as is possible by using the potential energy supplied free by gravity. This is another example of the view that pollutants cost money. Since the construction of the 'ring main' there has been a significant increase in the dividend paid to shareholders as well as possible reductions in the rate of increase in cost of water.

5.7 WATER TREATMENT

The conversion of water from natural sources into potable water is a fairly simple three stage process.

1 The removal of solid sedimentary and organic material is especially important when the source is a river or reservoir. It is generally achieved by passing the water through a sand filter bed. Aquifer water contains practically no sediment or suspended organic material, but it does often contain bacterial micro-organisms.
2 At this stage water from all sources will contain bacteria. Chlorine gas is used for sterilization, or disinfection.

3 There is some debate as to the effect of the chlorine, but it does leave an unpleasant taste in the water so the final process is dechlorination. This is achieved by passing sulphur dioxide through the water leaving a residual of free chlorine in the treated water to act on any bacteria that gets into the mains. From there it is pumped to the reservoir.

5.8 WATER USES

5.8.1 Demand

As living standards have improved so has the demand for potable water. In 1905 each person used, on average, 120 l per day. By the year 2000 this will have risen to around 450 l per day.

The demand for potable water fluctuates throughout the day. Generally the demand for water from midnight to around 3pm is below average for the day and from then until midnight above average. The production of potable water is never at the same rate as maximum consumption. The volume of the resource therefore depletes during the day and builds up again at night.

Annual rainfall in Britain varies with season. The summer months are on average driest. However, the extent of the difference changes with parts of the country. It is suggested that climatic warming will change weather patterns slightly so that we experience very wet winters and droughts will become common in the summer months. Where we can be more sure is in the distribution of rainfall. Most of the wet weather we experience comes via frontal systems from the west. As the air is forced over the hills and mountains rainfall occurs. It is no surprise to find that East Anglia receives approximately half the rainfall of Scotland or Wales.

If the overall demand for each region is compared with the quantity available from rainfall a similar picture emerges to that of the hard and soft rocks. There exists a plentiful supply through precipitation and relatively little demand in northern and western Britain tending to a greater demand with less supply in the southeast.

5.8.2 Industry

Electricity generation accounts for 36 per cent of all water abstracted. It is used as a coolant and then returned to the river. The water is taken directly from the river to the power station. Its only function is as a coolant so apart from screening it requires none of the treatment described in Section 5.7. Provided that the temperature of the river is not increased sufficiently to damage the flora and fauna, the process is of less importance to the environment than many others.

Other industries account for 11 per cent of abstractions.

5.8.3 Agriculture

Agriculture accounts for only 1 per cent of water abstracted, but this amount is increasing rapidly – 125 per cent since 1980 – and is dependent on weather. In drought conditions farmers using spray irrigation would increase demand considerably.

5.8.4 Leisure

Britain was for many years, a sea going nation. In the past it has boasted the world's largest navy and a huge fishing industry. Urban development has therefore tended to be concentrated near the coast. For many holiday makers a trip to the seaside is essential. Towns and villages close to sandy beaches experience dramatic increases in population for a few weeks of the year. Poole in Dorset experiences a 50 per cent increase in consumption during the holiday season. Other smaller holiday towns suffer greater increases.

5.8.5 Domestic

Piped mains water accounts for 52 per cent of the water abstracted from surface and ground water.

Accurate estimates of the amount of potable water used around Britain vary because of different needs of people in different regions. However the average household in Britain uses approximately half a tonne (around 500 l) of water each day. The figure for individuals is 110–150 l per day. This obviously varies with age group, i.e. young children require considerably more water than adults, and location, families in the south of England use more than those in Scotland. Table 5.1 shows how much water is required each time an appliance is used and compares

Table 5.1 Domestic water consumption

	Typical consumption of appliances in litres
Washing machine	100
Bath	80
Dishwasher	50
Shower	30
WC cistern	10

it with industrial use. What is not evident is the overall quantities used. For instance a family with three small children would use the washing machine at least once each day.

Some water engineers would separate the two types of use. The first is a demand for clean water. We drink it, wash our faces in it and cook in it. It is for these reasons that we demand and produce potable water. However, we also flush gallons down the lavatory, clean cars, and water the garden with it. When we look at domestic uses we find that the majority of the water we use does not need to be treated. It is proposed that, if the cost of treating water continues to increase we will, in the not too distant future, find building companies installing dual pipe systems in new properties. However the greatest cost to many water companies is the energy used in pumping the water to the reservoirs. A duplicate pipe system would double the pumping cost and involve the laying of a new main system. Perhaps the only way forward is to reduce consumption.

5.8.6 Metering

Each day in Britain we use around 20 million cubic metres of water of which a quarter is metered. As a general rule metering is used at commercial and temporary outlets such as building sites. Domestic users are less likely to be metered although experiments are taking place throughout the country. It is perhaps not suprising that the majority of these experiments are taking place in southern England because the water companies there have less rainfall than other areas and often a high population density.

5.9 DISTRIBUTION

Water distribution is the methods and plant used to transport potable water from the treatment works to the consumer. It is necessary for the water to be stored above the consumer so that there is sufficient pressure in the pipes to make the system work. The pressure available to the consumer is proportional to the difference between the the altitude of the water level in the reservoir and of the outlet (Figure 5.10). This is known as the **static head**.

In Figure 5.10 are three consumers. At the farm on the side of the hill the static head is 30 m. At the house in the valley it is 70 m and in the top flat nearby it is only 15 m. Where consumers are situated at almost the same altitude as the service reservoir extra pressure is introduced by constructing water towers

(Figure 5.11). Generally, they only serve properties with an altitude of which are less than 30 m below the service reservoir.

The greatest general need for water pressure comes from the fire service. The pressure required to propel water into the air is provided by a static head of 30 m at a hydrant served by a 75 mm (minimum) pipe. While the water undertakers (private water companies) are legally responsible for the supply of hydrants there is no stipulated pipe dimensions.

We should also consider the effect of too much pressure. Plumbing fittings are designed for use at moderate pressure. Higher pressure increases wear and tear of components such as taps, ball valves and washing machine parts and increases the chances of 'hammering', a loud banging caused by the sudden opening of a valve or tap under high pressure. The water supplier is also keen to keep the pressure down since the distribution mains often leak causing the loss of considerable volumes of treated water, and undermining the ground around the pipe. Obviously the volume lost is proportional to the static head.

Reduction in pressure is possible by either the introduction of pressure reducing valves, or by siting a second service reservoir 30 m below the first (as shown in Figure 5.10). If a high rise building is to be constructed such as is shown then water will be pumped to a storage cistern at the highest possible point to provide the required static head.

From the service reservoir the water flows through a network of pipes which reduce in diameter with distance. Typically the diameter of the pipe serving the grid is is 600 mm (24" on old plans). As water is used pressure is maintained by reducing the diameter so that at the farthest end of the distribution system the pipe diameter may be down to 100 mm (4").

From the main service pipe (Figure 5.12) potable water is carried into our buildings.

The quantity of water piped from the treatment works is often considerably less than that which is delivered to customers. Nationally the figure for undelivered water is around one third of the potable water produced. It is also used for sewer cleaning, fire fighting and mains testing etc. However the majority of water not delivered is lost through mains leakage.

5.10 SEWERAGE

The earliest sewers in Britain were used to remove storm water. Although the need for an effective method of removal of all human waste was recorded as far back as 1531 no authority existed then to

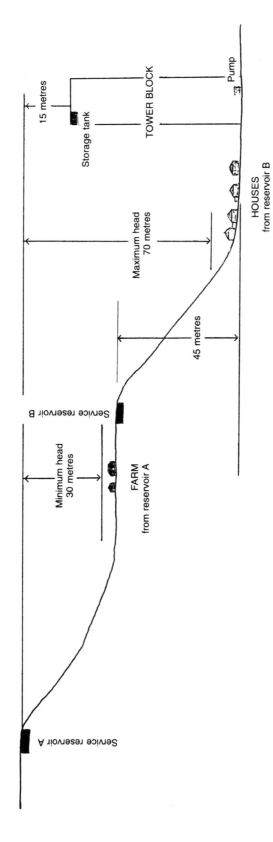

Figure 5.10 Static head

Figure 5.11 Typical water tower

coordinate its construction. Early in the seventeeth century the filthy conditions in areas of high population density were often attributed to piles of rotting garbage and overflowing cesspools. In the same way that the air pollution disaster of the early 1950s was responsible for succeeding legislation such as the Clean Air Act, the changes in attitude towards sanitation were primarily in response to the cholera epidemics of 1832, 1849, 1852, 1853 and 1854. In 1855 a sewage act was passed to improve the management of sewage so that every house was legally obliged to be connected to a sewer. The construction of the early sewers was started within ten years of the Act. By the beginning of the twentieth century the sewers had been taken over by the local authority. They increased the network and introduced a treatment system where the effluent was discharged into the river Thames and the solids dumped at sea.

The increased demand for water necessarily leads to a need for more, and better, sewers and drains, and a system of sewage treatment which is acceptable to both the built and natural environments. The major contaminant of coastal water in the UK is raw sewage and sewage effluent. EC targets for cleanliness have resulted in the construction of sewage treatment works capable of dealing with the maximum rather than the average population.

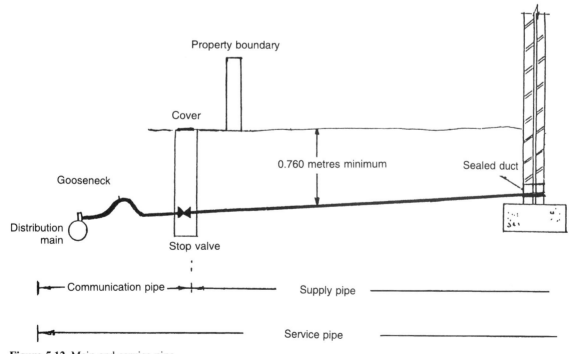

Figure 5.12 Main and service pipe

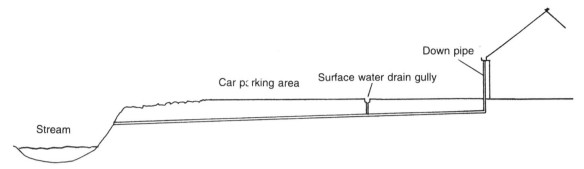

Figure 5.13 Surface water into a stream

5.10.1 Surface water

The water which falls on our buildings, car parks, pavements and roads is transported as **surface water**. It is considered relatively clean. There are three methods of disposal.

1 Disposal directly into a local stream. Although a cheap and effective method for the removal of surface water problems do arise when it is collected from paved or concreted areas. Accidental spills of liquids which pollute rivers, for example milk, are common, as is the intentional disposal of motor oil into surface water gullies.
2 In some areas the water undertakers prefer to remove surface water via a drain and from there to a surface water sewer which finally discharges into the sea or river.
3 If there is no suitable stream or nearby local authority surface water drain or sewer, surface water may be disposed of into a soakaway (Figure 5.14). The physical nature of the soil will determine the construction method.

5.10.2 Foul

The liquid and solid waste from the sanitary fittings inside of our buildings requires treatment before it can be returned to the natural water cycle and must be transported as **foul** to a sewage treatment plant. Internal waste pipes which convey the fluids from baths, basins or sinks are either 32 mm or 38 mm diameter. The pipe from the pan and the underground drainage and soil and vent pipe (svp) are usually 100 mm.

The disposal of foul from the site is, in the vast majority of cases, via a sewer to the water undertakers sewage treatment works. The only alternative for those some distance from the nearest sewer is the septic tank.

Leakage from sewer pipes is not as serious a problem as it may seem. The pipes are usually beneath a metre or more of saturated soil so that the water pressure outside the pipe is greater than the air pressure inside. Unlike water supply pipes, sewers tend to leak inwards (unless the pipe is actually broken, in which case it is a health hazard and should be replaced).

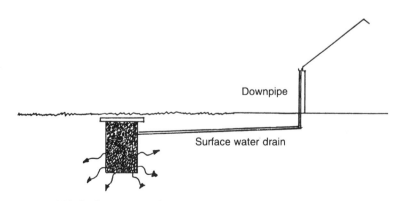

Figure 5.14 Surface water soakaway

5.10.3 Sewage systems

There are three different methods for dealing with the foul and surface water sewers.

The **separate system** consists of independent drains and sewers. The foul sewer conveys its waste to the sewage treatment plant while the surface water discharges into the sea or a river.

The **partially separate system** allows some of the surface water to mix with the foul. It has the advantage of diluting the sewage, and in wet weather the pipes are cleaned. The disadvantages are:

- that a two pipe system is required, although the remaining surface water pipe need only go to the nearest stream or other natural outlet
- the capacity of the sewage works must be sufficient to cope with increased volume of sewage, and the waste from surfaces.

The **combined system** is more common now. It allows for both drains to discharge into a combined sewer which has two advantages:

- it is obviously more economical
- the pipes are cleaned by even moderate rainfall.

The major disadvantage is that the sewage treatment works must be capable of not only dealing with very large volumes of liquid, but also the grit which is washed from the roads during storms. Most sewage works have a system which allows for the storage of a significant quantity of raw sewage in storm water tanks. When the storms become severe the incoming sewage is stored and then treated when the flow returns to normal. Very occasionally the flow of sewage is so great that it exceeds the capacity of the treatment section and is passed to the outfall. This is not as bad as it sounds. The volume of rainfall required for this to happen is so great that the polluting sewage component is very dilute.

The classification of a pipe as either a sewer or a drain is not entirely straightforward. However, as a broad generalization the waste leaves buildings in drains. When the drains connect and become the responsibility of the local authority, they become sewers. The function of the sewer is to carry sewage to an outlet such as a treatment works, river, estuary or open sea. During the journey the raw sewage undergoes aerobic degradation using oxygen from the air in the pipe. If it becomes starved of oxygen the biological breakdown becomes anaerobic and hydrogen sulphide is produced which is corrosive, a health hazard and can cause severe odour problems. The problem is minimized by ventilating the sewer pipe, and ensure that the flow is as rapid as is possible thus minimizing the time the sewage is in the pipe.

5.11 SEWAGE TREATMENT

Natural water contains dissolved oxygen from the atmosphere and from the photosynthesis of aquatic plants. It is used in respiration by animals, and by the decomposition of organic material. This creates an oxygen demand. In a healthy aquatic environment the rate of replenishment sustains a dissolved oxygen concentration at a level sufficient to support life in the water. Sewage contains a considerably greater concentration of biodegradable organic matter than does fresh water. The oxygen demand is therefore greater. To assess the condition of any sewage or effluent the demand is measured over five days. The dissolved oxygen is measured on the first day and the sample is then incubated at 20°C for five days when it is measured again. The difference is the **Biochemical Oxygen Demand** (BOD). The primary function of a sewage treatment works is to reduce the BOD to less than $20\,\mathrm{g\,m^{-3}}$.

5.11.1 Primary treatment – 1

On entering the treatment works the raw sewage is screened. Typically this process involves the use of two or more lifting screens which rotate in the flow of sewage lifting out floating solids. The screens (Figure 5.15) have a tooth gap of 25 mm. The second screen has a tooth gap of 6 mm. The screenings, a particularly unpleasant mixture of the byproducts of human activity, usually pass into a dewatering unit then into a skip for removal to a landfill site or incinerator. The screened sewage is then passed into settlement tanks where the coarse sand and grit sinks to the bottom and is raked off.

5.11.2 Primary treatment – 2

From there the sewage goes to the sedimentation tanks which are sufficiently large to produce the very low velocities necessary for the settlement of fine solids. Some of the solids removed at this stage are of an organic nature. The removal of biodegradable material reduces the BOD. There are several methods for removal of the sludge. The most common is to incorporate scrapers into the settling tank (Figure 5.16) which rotates slowly moving the sludge, with a little help from gravity, towards the sludge well. Floating material, or scum, is either

Figure 5.15 Screens

(a)

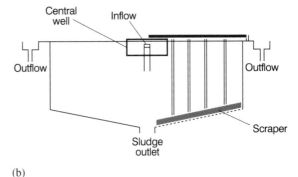

(b)

Figure 5.16 Settling tank

scraped from the surface by a blade on the on the scraper arm, or is sprayed with a chemical which causes it to sink.

While the designs of primary settling tanks vary the basic operations are the same.

1 The velocity and turbulence across the tank should be low. The best place for the inlet is therefore in the middle of the tank and below the surface.
2 The outlet should be as high as possible. The function of the tank suggests that material will be moving under gravity so the cleanest sewage will be just below the surface.
3 The peripheral weir should be just below the surface and protected by a scum board so that floating material cannot pass out of the settling tank.
4 The sludge collects in the base of the tank by a combination of the rotating scraper and the acceleration due to gravity. From there it is pumped to the dewatering plant. Anaerobic bacteria then work on the sludge and break it down further leaving a fibre like material which is then disposed of by either landfill or incineration.

5.11.3 Secondary treatment

Sewage enterning this stage typically has a BOD of around $175\,\mathrm{g\,m^{-3}}$. The next step is to increase the concentration of disolved oxygen in the sewage.

Oxidation may be achieved by pumping air through **diffusers** through the sewage until it reaches a predetermined dissolved oxygen concentration. As the dissolved oxygen rises a computer relays information back to the pumps.

5.11.3.1 Biological filtering

A more common method however is biological oxidation. This is achieve by passing the sewage over a medium which has a large surface area and is well ventilated. The best results are achieved using clinker or blast furnace slag, though gravel and crushed rock are also used. Artificial filter material is available but generally at a higher costs and so far they tend to be less efficient in the long term. The micro-organisms, mostly bacteria, live on the surface of the medium. They require a steady source of food, the organic material in the sewage, and a supply of oxygen from the atmosphere. Unfortunately the bacteria are not capable of digesting every sort of unwanted substance. Toxic substances and some organic material they are unaccustomed to (such as milk) can damage the filter bed. Given a plentiful supply of food and air these bacteria will continue to grow until the filter system becomes blocked, so the presence of small animals, especially the larvae of flies is beneficial to the system. In the final settling tank the humus is removed leaving a BOD of approximately $20 \, \mathrm{g \, m^{-3}}$ and suspended solids of around $30 \, \mathrm{g \, m^{-3}}$.

5.11.4 Activated sludge

An alternative, and effective method of increasing the oxygen concentration in the sewage is to agitate the sewage from the primary settling tanks in an aeration plant. The introduction of air into the sewage encourages the growth of aerobic bacteria. When the sewage passes into the final settling tank, which is of similar design to the sedimentation tanks, the activated sludge is removed. Since it contains oxygen and aerobic bacteria some of it is combined with the sewage from the primary settling tank to increase the efficiency of the process. The sludge that is not returned it treated and disposed of along with the primary sludge. It is estimated by the Government Statistical Service that one million tonnes (dry mass) of sewage sludge is produced annually.

The activated sludge method has some advantages over the filter method. They are:

1 the activated sludge method requires far less land
2 greater control especially with computers
3 less pipework and pumps.

Biological filters on the other hand require less maintenance, are less susceptible to toxic substances, require less energy and are less noisy. The running costs are about the same but a greater proportion of the activated sludge method comes from the cost of the energy used in the aeration plant, so increases in energy costs may tip the balance slightly in one direction while land values could have the opposite effect.

5.11.5 Tertiary treatment

The effluent from the final settling tank should have a BOD of not more than $20 \, \mathrm{g \, m^{-3}}$, and a suspended solid of not more than $30 \, \mathrm{g \, m^{-3}}$, and this is generally the case. In a few cases these readings must be lower. For instance if the river were used as a source downstream, or if the discharge from the works were greater than one eighth of the dry weather discharge of the river. In these cases further, or tertiary treatment is required. This may involve:

1 running the effluent over grassland
2 storage in large lagoons
3 the use of sand filters similar to those used in water treatment.

In 1991 The Urban Waste and Water Treatment Directive from the Department of the Environment (DOE) set priorities for the discharge of treated sewage or effluent. The directive deals with the quantity discharged and the sensitivity of the receiving river or stream. It lays down that primary treatment be the norm in areas where the effluent is dispersed rapidly by natural processes. Secondary treatment should be the minimum elsewhere. In areas where the natural dispersal is low, or sensitive areas, more stringent treatment is necessary.

Overall this suggests an attempt to cut down the pollution of water and will represent a sizable increase in orders for the construction industry.

5.12 EXAMPLES

5.12.1 Local case study

At the begining of this chapter we suggested that an all encompassing description of the artificial water cycle was all but impossible. A useful addition would therefore be to compare the general system described here to the local system in a case study.

6

The internal environment

6.1 INTRODUCTION

When we move around in the open our expectations are that the temperature will change hour to hour and month to month, the amount of light will vary through the day at times making it difficult to see properly. We accept that it is difficult to see properly once the sun begins to set. The air moves around us, even to the extent of causing responsive movements. We enjoy the occasional summer breeze. To overcome the lower temperatures we insulate our bodies with layers of clothes.

When we move into the internal environment our expectations change. Building designers and construction services engineers are responsible for providing us with natural light during the day and artificial light at night. Tolerance of temperature change is reduced inside a building: a change of a few degrees is, to some, the cause of irritation and discomfort.

We expect air to be free from unpleasant odours and pollutants. Sound insulation should be sufficient to isolate one home or office from another and from the internal to the external environment and vice versa.

6.2 SPACE

An important feature of a well designed domestic property is the provision of room sizes (determined by floor area) suitable for the economic, social and in some cases religious status of the occupants. Space in the home is a measure of social status, in the same way as the amount of land with the property and the type of car parked in the drive. Large rooms therefore bring with them a 'feel good factor'. This explains the tendency for owners of terraced homes with two living rooms to demolish the partition wall thus producing one large room. The apparent increase in space is at the expense of privacy and increased fuel bills (especially in rooms with a large window at each end).

While the size of the living area has implications for fuel use and social standing, the more important consideration is overcrowding. Overcrowding is associated with poverty, and can occur in properties of any size. If indoor space levels are low then parents may have to share a bedroom with children, teenage children of different sexes may share bedrooms, young married couples may be forced to share the kitchen with parents. The consequence of overcrowding is a decrease in the quality of the internal environment. Infectious diseases are rapidly transmitted, air pollution is increased and, in the UK especially, condensation may become severe. The difficulties created by a lack of space are not individually severe, but often combine to create an oppressive environment.

Following the onset of the industrial revolution (Chapter 1) people moved into overcrowded accommodation close to their jobs. Common causes of death then were consumption (tuberculosis) and pneumonia, both respiratory illnesses.

Modern studies have shown that children from crowded homes on average are more likely to:

1 miss schooling because of respiratory illness
2 be slightly shorter
3 score lower marks in class tests
4 get less sleep
5 be more susceptible to accidents
6 show evidence of mild mental disorders when mature
7 suffer from family violence or sexual problems.

The importance of space in the home has changed the way houses are described. A large three-bedroomed house may be better described as a five person house, as this infers a design limit to the number of occupants. In the public sector there is no doubt that limits would, in some cases, improve the health of the occupants. However allowances must be made for the difficulties involved in rehousing growing families.

6.3 UNWANTED MOISTURE

Dampness in the home is thought to either cause or aggravate rheumatism, arthritis and respiratory illness such as bronchitis and pneumonia. There is an obvious risk to those who are susceptible to such illnesses. Most at risk are the elderly, those with on-going respiratory problems such as asthma, and the very young who may suffer for the rest of their lives as a result of the inhalation of the micro-organisms.

The mould associated with dampness will also destroy furniture, clothes and wall coverings. Because water is a better conductor of heat than is air the house also feels colder. This can result in higher fuel bills and depression.

6.3.1 Condensation

Here the source of the moisture is from inside the building rather than from the outside. The extent of condensation is primarily determined by temperature and humidity. **Absolute humidity** is the amount of water vapour in air measured in grams per cubic metre (Figure 6.1). **Specific humidity** is similar but measured in grams of moisture per kilogram of air.

Figure 6.1 shows how humidity also depends on the temperature of the air. Cold air is not capable of holding as much water vapour as warm air so we need a measure of the percentage of the saturation value. **Relative humidity** (RH) is the amount of water vapour in air at a specific temperature as a percentage. Air is saturated at 100 per cent, moist or damp from 95–99 per cent, and dry at 50 per cent. The lowest RH ever recorded was 10 per cent over a hot desert.

Question 6.1

If the relative humidity is 100 per cent in the morning at 5°C and the temperature increases to 20°C by lunchtime, what is the new relative humidity? (Assume no change in vapour content.)

Answer
From Figure 6.1 if the relative humidity is 100 per cent and the temperature 5°C the air will hold 7 g of water vapour. At 20°C it could hold 17.1 g. Relative humidity is the amount of water vapour as a percentage of maximum quantity it could support. So we need to find out what percentage of 17.1 is 7.

$$\frac{7}{17.1} \times 100 = 40.9 \text{ per cent}$$

As the temperature in a room rises the air absorbs moisture from many sources including our bodies,

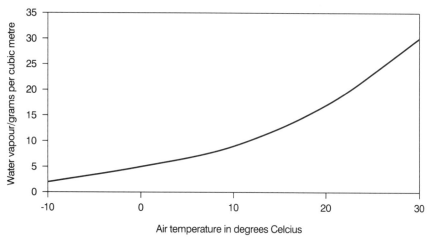

Figure 6.1 Humidity at saturation and temperature

plants and fish tanks. When that air comes in contact with a cold surface, for instance the glass in the window, a temperature is reached where the air is saturated. This is the **dew point**.

A common complaint from people who have sealed unit replacement double glazing, or good quality secondary glazing fitted to their homes is the increase in condensation. This is especially the case where the initial attraction to the product is a reduction in the fuel bill created by the lower U-value of the glazed unit and reduced air changes resulting from the installation of draught excluders. If no change is made in the way the room is heated the temperature will be constantly higher than the occupants were used to. The RH will in, many cases, exceed 100 per cent. When the people leave the room at night the temperature falls more slowly than it did before the room was sealed but because of the super saturation condensation appears almost immediately and can be severe, even resulting in water running down the walls. There are many sources of water vapour. Four people produce 5 l of water per day and for approximately one third of that time they are in a poorly ventilated bedroom. If the bedroom is small the concentration of water in the air will be greater.

The average household with a gas supply produces approximately 4 l of water vapour per day simply by burning gas in air. Further moisture is added by washing dishes, washing floors, showering and bathing and watering house plants.

Question 6.2

The air temperature in a lounge is 22°C. The air is saturated, (RH = 100 per cent). If the room is 5 m × 4 m and 2.4 m high, how much water will condense if the temperature falls by 8°C overnight? Assume no air loss.

Answer
From the graph (Figure 6.1) we can see that at 22°C and 100 per cent saturation each m³ of air contains 18.5 g of water. At 14°C and 100 percent humidity each m³ will contain 12.5 g of water. The volume of the room is:

$$4 \times 4 \times 2.4 \text{ m}^3 = 38.4 \text{ m}^3$$

The mass of water is therefore:

$$12.5 \text{ g m}^{-3} \times 38.4 \text{ m}^3 = 480 \text{ g}$$

Question 6.3

A bedroom (4 m × 3 m × 2.4 m) is heated to 25°C. Humidity is 120 per cent. The temperature falls to 10°C overnight. How much water vapour is produced?

Answer
Saturation at 25°C is 23 g m^{-3} so if humidity is 120 per cent absolute humidity will be:

$$23 \times \frac{120}{100} = 27.6 \text{ g}^{-3}$$

At 10°C the air is saturated at 9.4 g m^{-3} so each cubic meter will produce 27.6 − 9.4 = 18.2 g of condensation. The volume of the room is 28.8 m³, so the total condensation produced is:

$$28.8 \text{ m}^3 \times 18.2 \text{ gm}^{-3} = \underline{524.16 \text{ g}}$$

Initially this may not cause a problem but if it goes unchecked it may cause the same damage as other forms of dampness. Mould from condensation forms in the colder areas of the room and where there is less air movement. A typical example would have a thin patchy layer above the skirting board with spots or patches in corners or behind furniture. Mould feeds on the starch in wallpapers, paints, clothes and needs water to survive.

In excessive cases condensation dampness is often confused with rising damp. The major differences are that rising damp leaves a white salt on the wall when dry, and condensation mould is blotchy.

The healthiest relative humidity for humans is around 50 per cent and for moulds and fungi a RH of 70 per cent is required for at least 12 hours. Populations of house dust mites (also capable of causing respiratory problems) will not grow unless the RH is between 75–80 per cent.

Unlike rising or penetrating damp, condensation cannot be remedied by a repair to the building. However, it is possible to design buildings which are less susceptible to condensation. Condensation mould is therefore the responsibility of the builder and the occupant. The most useful actions the occupant could take would include:

1 reduce water production by non essentials such as house plants and indoor fish tanks
2 keep the kitchen and bathroom doors closed when steam is being produced from a sink or a bath

3 increase the ventilation in rooms where mould grows

4 keep the temperature of the walls close to that of the air as mould seldom grows on internal partition walls

5 only use free standing gas or oil burners when absolutely necessary.

There are implications in the above list for building designers.

1 A low wall U-value can inhibit the growth of mould by reducing the temperature difference between the air and the wall surface.

2 Trickle ventilators should be installed in sealed unit windows especially in bedrooms.

3 One of the major contributing factors to dampness is a constantly changing temperature. The running cost of the heating system is therefore considerably more important than the installation cost.

4 The heating system in a house should negate the use of mobile self contained heaters which produce water vapour by burning fuels such as propane or paraffin.

6.3.2 Rising damp

Rising damp is caused by the failure or absence of a damp proof course (d.p.c.). The moisture below the dpc moves up the wall by capillarity and brings with it in solution salts from the structure or ground below. The most obvious evidence is a usually large area of dampness which appears above the skirting board. On dryer days the water evaporates and leaves a white patch of salt.

6.3.3 Penetrating dampness

Penetrating dampness is the result of either a permeable material, such as, mortar bridging the cavity, or a solid wall which has become permeable with age. Penetrating dampness usually appears as a damp patch on the internal wall. It is often found below windows where it is caused by a short vertical damp proof membrane fitted to the side of the window frame. However, dampness in that area may also be caused by condensation from the window dripping from the window board at the corners thus staining the wall. If the dampness is the result of permeable external brickwork a water-proofing liquid which is absorbed by the wall will improve the situation. In many cases it is the result of cracks and gaps in the structure, especially around frames.

6.4 LIGHT

6.4.1 Natural light

In hotter countries, and for a few days each year in the UK, control of incoming sunlight is necessary in the office, factory and home. Excessive sunlight is unpleasant and can cause illness to workers ranging from short term nausea typical of an employee sitting with the sun shining on their back, or sunburn and its associated problems suffered by those working in the open. In factories and offices the effect of direct sunlight on machinery can cause a variety of problems. Engineering machinery used in the production of components to fine tolerances are unlikely to perform successfully if they intermittently expand and contract as a result of direct radiation. A design for the roofs of industrial buildings when diffuse light was required but direct light unwanted, is the northern light (Figure 6.2). The sun rising in the east makes an arc in the sky to the south and sets in the west. A window facing north will therefore never receive direct sunlight.

In the home natural light creates a visible connection between the natural and built environments. Houses with wide patio doors through which the semi-natural world of the garden is easily visible are generally perceived as more desirable than those with small windows. Penthouse apartments have large windows looking out into the natural world – prisons have small windows where only the artificial world is visible. Furthermore, larger windows allow more light into the room which increases the temperature slightly during daylight hours. It would seem then that a domestic property will sell faster for a higher price if it has large windows overlooking a natural landscape. There are, however, problems. Heat is lost through the walls and windows at a different rate. The U-value of a cavity wall in a modern property will be around $0.4 \, \text{W/m}^2\text{K}$ depending on the construction method. The U-value of a double glazed sealed unit consisting of two sheets of 4 mm glass and a 20 mm air gap is $2.8 \, \text{W/m}^2\text{K}$. Because of recent legislation governing

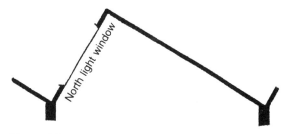

Figure 6.2 Northern light window

the thermal efficiency of domestic properties, and the quite large differences in U-value between walls and windows the size and number of windows in the design is limited.

Solar radiation also increases the temperature of the floor, walls and furniture, but on hot summer days it is necessary to reduce the radiation with curtains or blinds because the sunlight can damage or discolour furniture. Reputable door and window manufacturers will be able to offer glass which cuts out the damaging light without spoiling the effect of the window.

6.4.2 Artificial light

The ordinary household light bulb works by the heating of a tungsten filament so that it glows brightly. The intensity of the light emitted depends on:

1 the power of the bulb
2 the use the room is put to
3 the requirements of individuals.

Offices and workshops occupied by many people are often illuminated by fluorescent tubes which flicker at predetermined frequencies. To the human eye a flicker of greater than 50 Hz is undetectable. It seems continuous. However, recent studies have shown that headaches, especially those which are associated with the eyes, are caused by flickers of specific frequencies, notably 100 Hz. The frequency of flicker in the tube is produced by the ballast. A simple remedy is therefore to change the ballast for one which produces a different frequency.

6.5 THERMAL COMFORT

The design of buildings which are capable of satisfying the thermal comfort criteria of all those likely to use them is a complex process impaired by the wide variation in the thermal requirements of individuals. The features which influence each individual's thermal comfort are either personal or external. The external influences on the immediate environment include air temperature and movement. Personal influences depends upon gender, metabolic rate and the clothes we wear.

6.5.1 Air temperature

The environmental temperature is often used for design purposes. The purpose of the room or building determines the design temperature. Generally they fall between 16°C for a factory, through 18°C for an office, to 21°C for a living room.

6.5.2 Heat output

The heat produced by our bodies is usually trapped by our clothing and to a lesser extent body hair. Generally the more heat we produce the warmer we feel. Two factors control our heat output. The first is determined by our metabolic rate. This indicates the rate at which our bodies produce heat as a result of its normal functions. If a large person and a small person have the same metabolic rate the surface temperature of the large person would be lower because of the difference in surface area. Similarly two people of the same size could have different metabolic rates and therefore produce energy at different rates.

The second factor which governs heat output is activity. A hod carrier working flat out will produce heat at around 500 W (Watts = Joules per second). An office worker produces 130–150 W, and a sleeping person around 70 W. Heat output, and the ability to keep warm diminishes with age.

6.5.3 Clothing

The restriction of the movement of air by clothes has an insulating effect. Clothes which trap little or no air, for instance silk shirts and tights, influence our surface body temperatures less than those which trap relatively large quantities, like a woollen pullover.

6.5.4 Air movement

The movement of air in an enclosed space (not to be confused with ventilation) is caused by either people moving around, or temperature differences at different parts of a room. The latter is especially noticeable in a room where the air is otherwise still. A good example is a lounge with a window at one end, and patio door or french window at the other. The air in the room is heated. The air close to the windows is cooled so that it drops to the floor to be replaced by heated air from the room thus producing a circulatory effect and the movement of colder air across the floor as shown in (Figure 6.3). This is often mistaken for a draught entering the room from beneath the door and followed by an ineffective remedy.

If the air speed in a room increased by only $0.1\,\mathrm{ms^{-1}}$ a temperature increase of 3°C is required to achieve the same level of thermal comfort.

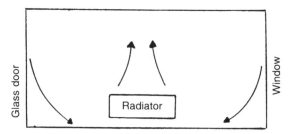

Figure 6.3 Air circulation caused by temperature differences

6.5.5 Gender

As a general rule typical comfort temperatures are slightly higher for women than for men. Women are also generally less tolerant of fluctuations.

6.5.6 Extreme heat

In the British Isles the temperature exceeds 30°C so rarely that it is usually considered a benefit or at worst a moderate discomfort since we accept that it is usually a short lived phenomena. While insulation usually serves to keep in heat and therefore save money, it will also prevents excess heat entering the building. This is especially relevant with single leaf construction where the outer cladding is heated by radiation as anyone who has been into a loft on a hot day will testify. Similarly properties constructed to take maximum advantage of insolation (the radiation which reaches the surface of the earth) tend to overheat in the hot summer days. It is therefore important when designing buildings to consider the orientation of windows and fully glazed doors if we are to take advantage of natural heat and light without causing overheating problems. South facing windows may be fitted with blinds, either internally or externally, to reduce excessive radiation. It is always a good idea when a south facing window is selected to include an opening area as close to the ceiling as is possible thus allowing the escape of the hottest air in the room in the summer and its retention in the winter. In more affluent areas of the world where long hot summers are the norm properties have been constructed with air conditioning systems. In Britain we are more able to take advantage of natural light, shade and ventilation thus generally precluding the necessity for another energy consuming device.

6.5.7 Extreme cold

People who are working in uncomfortably cold conditions do not work as efficiently as they other-wise would. Extreme cold can cause great discomfort, illness, or even death through hypothermia. The elderly especially are intolerant of extreme cold and those who rely solely on state pensions are often unable to afford to rectify the situation by insulating and heating their homes. Public costs associated with a lack of thermal insulation include excessive use of fossil fuels which increases pollution and reduces the reserve, and the cost to the state of care for the elderly as a result of illnesses.

6.6 HEAT BUDGET

Heat is gained and lost from buildings. If the internal temperature is to be maintained the total heat output must be matched by the input. A building may gain heat as a result of the natural environment such as sunlight on a window, or by artificial means such as a central heating system. In temperate climates the natural heat gains are for the majority of the year less than the natural heat losses. When the resulting reduction in temperature adversely affects comfort the difference is made up by purchasing heat in the form of fuel.

6.6.1 Natural factors

6.6.1.1 Season and latitude

The daylight hours in mid winter vary from north to south across the British Isles. The north coast of Scotland receives around six hours in mid winter. The south coast of England receives around nine hours. In the summer the opposite is true. The south coast receiving fewer hours of daylight than Scotland. Solar radiation reaching the surface of the earth, insolation, is therefore governed by both the latitude of the site and the time of year.

6.6.1.2 Weather conditions

The average intensity of the insolation in Britain is also determined by the cloud cover. At any one site then the quantity of heat gained is perhaps pot luck. However, designers considering incorporating solar gain features might wish to consider the expected cloud cover for particular regions of the country. Our weather systems generally approach the country from the west. As the air is lifted over the mountains in the west of Britain cloud forms. Annual rainfall figures show clearly that the east of the country is generally flatter, lower and dryer than the west. The wettest towns in Britain are situated to the west of mountain ranges. Fort William, for instance is situated on the western side of Ben Nevis. Manchester is famous for its rainfall and is situated to the west of the Pennines. In contrast, Sussex and Kent are as dry as some areas around the Mediterranean.

6.6.2 Design factors

The maximum rate of solar gain is achieved through windows and doors so it practical to design the layout of the site and the buildings so that the windows are appropriately directed. The notion that all large windows should face south is an over simplification of the situation. The primary function of windows is one of illumination and not heating. Direct illumination in the work-place may increase the temperature and therefore save money, but direct radiation is often the cause of discomfort. This was overcome in early factory design by building a northern light window (Figure 6.3) into the roof of the building thus allowing diffuse light to enter the building without the disadvantage of direct radiation.

The angle of incident of the sun's rays on the window is also an important feature. When the sun is high in the sky, at midday in June and July, much of the energy will not penetrate the glass. The maximum solar gain occurs through a window which is at a right angle to the path of the sun's rays. When the sun is just above the horizon the intensity of radiation is less but, because of the angle of incidence the gain is still significant. The rate of gain per unit time for a south facing window is, for the same reason, beneficial in the spring and autumn when the sun is lower in the sky. Solar gain through the walls of the building is complex, not least because of the variety of different external wall designs and materials. A traditional brick/block construction with a cavity fill insulation will receive radiation at the outer leaf and the temperature will increase. However, because of the density of the material and the insulation the heat flow towards the inside is very slow and unlikely to have significant effect before the sun goes down and the flow direction is reversed.

6.6.3 Non solar heat gains

Most of the activities that take place in the home and the office give off energy as a by-product of their design function.

Fluorescent and tungsten **lighting**, are the types most commonly used when illuminating to an acceptable standard, e.g. 400 lux in an office produce heat at the rate of around 30 W m^{-2} of floor. Although this is

a small amount some offices are of considerable size.

The amount of heat generated by **people** depends on their metabolic rates and activity. Estimates of heat emissions range from around 100–180 W.

Heat emitted from **equipment** tends to depend on their design function. Appliances which, as a part of their function, contain some form of heater obviously emit more heat than others. A television produces around 100 W, and a refrigerator 150 W. A hair dryer on the other hand runs at around 800 W and a gas cooker 3000 W per burner. An office printer with a heater produces more than 1000 W and VDU more than 500 W.

In well insulated buildings these gains can be significant, more so in commercial buildings than domestic and should neither be wasted or ignored in the design of heating systems.

6.7 ENERGY USE IN BUILDINGS

Half of the UK energy produced is used by building services and half of that in domestic dwellings. The current impetus to provide energy efficient buildings is a combination of ever increasing price, and the cost to the environment.

There are three very strong arguments for the efficient use of energy resources.

1 The pollution from energy production is at present considered responsible for global warming and acid rain.
2 Fossil deposits, especially oil, are the raw materials used in the production of many materials, especially plastics. Many technologists believe that to continually deplete our reserves of coal and oil, or to dispense with them for political gain, is an unacceptable waste.
3 Energy costs can represent a significant proportion of domestic and business expenditure.

6.8 THERMAL INSULATION

The importance of insulation was emphasized when fuel prices increased rapidly during the 1970s. Until then the financial outlay involved in insulating homes and workplaces would not be considered recoverable in a reasonable time. Now insulation specialists use figures to show that their product will pay for itself in a few years.

The best thermal insulators are gases because the molecules are more widely spaced than in liquids or solids. However the molecules in gases are free to move and therefore transfer heat. This is overcome by using a material which traps small pockets of a gas, usually air. Inert gas is often used in sealed unit double glazing.

Fibreglass mats are flexible and easy to fit between the ceiling ties and can be cut without difficulty, though they do represent a minor health hazard. Fibreglass is also used as insulation in a traditional cavity wall, and between the studs in timber framed construction. Foamed polyurethane is used in the insulation of existing buildings (urea formaldehyde is discussed in Chapter 4). A slightly cheaper method of insulating an existing cavity is to use expanded polystyrene beads. There is no significant difference in thermal efficiency but care must be taken on two counts. First the beads are injected with a resin so that a solid mass is formed. If this is not carried out correctly any opening in the cavity, or later opening of the cavity will result in a loss of material. Secondly, electricians when rewiring a property will often drop a cable through the cavity rather than damage internal decorations. Recent evidence suggests that a reaction occurs between the sheathing of the cable and the polystyrene beads which results in the exposure of the wire.

A U-value denotes the rate at which heat energy is transmitted through an element of a constant thickness. (unit Watts/m^2 °C) If a material or element has a low U-value it is a good thermal insulator. If the U-value is high it is a good conductor of heat. The U-value of a single brick solid wall with 15 mm of plaster is around 2.3 W/m^2 °C. A cavity wall with two brick leaves and 15 mm of plaster is around 1.5 W/m^2 °C. Timber frame cavity walls with a cladding of facing bricks achieve a U-value of about 0.35 W/m^2 °C. A single glazed metal framed window has a U-value of about 5.6 W/m^2 °C, approximately twice that of a sealed unit with a 20 mm air gap.

6.9 SOUND INSULATION

For many years it has been accepted that sound insulation is best produced by placing dense material in the path of the soundwaves. However timber framed construction has showed clearly that an alternative method of party wall construction produces excellent levels of sound insulation by varying the material through which the sound travels.

6.10 VENTILATION

Modern building design requires that particles of dust and moisture are removed from the air prior to recirculation. In modern offices imported air is kept to a minimum because:

- heating or air conditioning costs
- as is often the case in cities, the outside air is heavily polluted.

In the 1950s most buildings were constructed with provision for an adequate supply of external fresh air. When commercial buildings are modernized the metal casement windows are often replaced with plastic sealed unit windows which do not open, and the building is made airtight. There have been many examples of ventilation which have either been blocked or switched off to increase comfort and save energy.

Homes are often fitted with sealed unit double glazing and draught excluders around openings in doors and windows. so that the same air is breathed for longer and therefore contains increased concentrations of pollutant and water vapour. It also results in a lack of dilution of odours, which may be irritating, or in the case of cigarette smoke and vapour from solvents, a long term danger.

If we seal our homes and offices, we must make provision for real ventilation, ie. internal air is diluted or replaced by external fresh air. Designers of the controlled internal environment will favour the automatic means such as electric extractor fans. In the home the answer is simpler – turn the heating down a little, wear more clothes and open the window a little for a few hours each day.

The quality of the air in an occupied space is affected by human activity. Respiration uses oxygen and produces carbon dioxide, heat and water vapour. At rest each breath takes in approximately half a litre of air which, if fresh, is 20.95 per cent oxygen and 0.04 per cent carbon dioxide. The expired air is around 16.5 per cent oxygen, a 21 per cent decrease, and 4 per cent carbon dioxide, a 1000 per cent increase. If oxygen intake is deficient the senses, especially sight are impaired. A slight deficiency spread over several hours is believed to result in the adoption of an uncharacteristically irresponsible attitude. More important though is the ten fold increase in carbon dioxide. If the air taken into the lung is rich in carbon dioxide the blood will become enriched, and respiration and pulse rates will increase. For confirmation of the danger involved in breathing air rich in CO_2 consider the extreme effects. When people who, either accidentally or intentionally, place a plastic bag over their heads, it is the increase in carbon dioxide which kills them, not the lack of oxygen.

Further use is made of ventilation in the control of humidity. If the relative humidity in offices falls below around 40 per cent static electricity builds up, staff suffer from dry throats and skin. At above 70 per cent the environment becomes oppressive at high temperatures and colder than it actually is at low temperatures.

As well as the human activities already described the use of correcting fluid, litho-offset printers, photocopiers and cleaning fluid all produce indoor air pollution. In the home contamination comes from the use of cigarettes, gas cookers and some cleaning fluids. The relative humidity can also be very high and is often saturated (over 100 per cent).

Finally, ventilation is necessary for the use of some fires and boilers. The pollutants from the combustion of hydrocarbon gases are very dangerous. The free standing propane burners which are often used as back up to a heating system are usually efficient and clean when running well. The burning of propane in air is explained by the following chemical reaction:

$$C_3H_8 + 5O_2 = 3CO_2 + 4H_2O$$

i.e. propane plus oxygen equals carbon dioxide plus water

which says that one molecule of propane burned in five molecules of oxygen produces three molecules of carbon dioxide and four molecules of water. The real problem is in the quantities. 1 kg of propane will take 3.6 kg of oxygen out of the air and replace it with 3 kg carbon dioxide and 1.6 kg of water.

For example, if a room is 3 m $\times$ 4 m and 2.4 m high, assume that there are no air changes, i.e. it is airtight, and that the original amount of CO_2 is negligible. If 1 kg of propane were burned what would be the resulting concentration of CO_2 in the air as a percentage? The density of air is about $2 \, kg \, m^{-3}$.

The mass of air in the room is found using:

mass = density $\times$ volume

   = $2 \, kg \, m^{-3} \times 4 \, m \times 3 \, m \times 2.4 \, m$

   = 57.6 kg

The combustion process uses 3.6 kg of oxygen which reduces the mass of air to 54 kg, and adds

1.6 kg of water vapour and 3 kg of carbon dioxide which increases the mass to 58.6 kg. The percentage of carbon dioxide can be found by:

$$\frac{\text{mass of } CO_2}{\text{mass of air}} \times 100 = \text{percentage } CO_2$$

$$\frac{3}{58.6} \times 100 = 5.1$$

The normal concentration of CO_2 in air is around 0.03 per cent. By what factor has it increased?

$$\frac{5.3\%}{0.03\%} = 177$$

This shows the need for ventilation where non-flued burners are used. The practice of cutting down the number of air changes per hour to save fuel must therefore be accompanied by a strict policy on the control of combustion gases by burner design and ventilation.

Of course this assumes that the burner is working perfectly. When the efficiency of fires is reduced by age or lack of servicing, incomplete combustion occurs and the desired reactions are partially replaced by others producing a more dangerous mixture of gases which includes carbon monoxide.

The incorporation of a heating system which discharges flue gases outside the building was once considered a luxury. It is now considered an essential aspect of building design.

6.11 BUILDING REGULATIONS AND ENERGY USE

The design and construction of new buildings, changes in use of existing buildings and some alterations, are controlled by building regulations as prescribed by the Building Act 1984. There is no provision in the act for environmental protection since existing legislation is considered adequate. The government (1995) has an ongoing policy of deregulation and is unlikely to introduce new powers of control. Some regulations exist which have, either by accident or intent, a direct influence on the environment. Building regulations controlling thermal insulation were introduced in 1979, and revised in 1985, 1990 and 1994 in response to the governments commitment to reduce carbon dioxide emissions to 1990 levels by the year 2000.

Regulations governing environmental considerations are often subsidiary to other concerns. In this case a test of cost-effectiveness may be applied. The general claw back time for insulating new houses to current (1995) insulation standards is 5–12 years. The apparent lack of accuracy is created by the diversity of construction materials available, construction methods and future fuel costs.

As with any new engineering methods there are technical changes which introduce unforeseen problems. One such problem is a result of the introduction of efficient cavity wall insulation. In winter the heat loss from properties increases the temperature of the external walls so the external wall is no longer heated from inside the property and its susceptibility to frost action is increased. Incorrectly installed impermeable cavity wall insulation may result in an increase in moisture penetration. The current failure rate is approximately 2 per cent. This cost of repair is sufficient to deter some builders from using the system.

Future changes in thermal insulation standards will make it practically impossible to achieve the required standard without cavity fill of some kind. It is therefore important that cavity wall insulation should be installed to the technical guidance issued by the Building Research Establishment. It has become evident from experience that the properties most in need of cavity wall insulation – those in exposed positions – are those which are most likely to suffer failure.

The 1990 regulations introduced the concept of the trade-off where a builder would increase the insulation in one component to a greater extent than is required by the standards so that another component could be omitted. Current evidence shows that many builders are opting for sealed unit double glazing on increased window areas which allows them to leave an air gap inside the cavity wall. It is unlikely that this will still be possible if insulation standards are changed again. Whether standards and regulations go too far is debatable. Many construction companies are concerned with the extra cost and technical difficulty and suggest that a better approach would be to deal with energy use in existing buildings, especially since the vast majority of buildings were constructed before any of these standards were in place.

6.12 INDOOR AIR POLLUTION

The air in buildings contains substances and micro-organisms that are potentially harmful to health. They originate from heating and ventilating systems, building materials, furnishing materials, cleaning products, dust in carpets and on shelves, fuel consumption and household activities.

6.12.1 Hydrocarbons

Pollutants from fuels include carbon dioxide (CO_2), Carbon monoxide (CO), and oxides of nitrogen (commonly NO_2). Nitrous oxides are formed when hydrocarbons are burned in air. At high temperatures the nitrogen and oxygen, which make up 99 per cent of the atmosphere, produce nitrous oxides. Nitrous oxide concentrations therefore increase by the use of internal combustion engines and gas cookers. Concentrations also increase in the home, when the external temperature is lower, as a result of a reduction in ventilation.

6.12.2 Micro-organisms

House dust mites feed on skin scales. Their numbers are rising possibly because of the increase in relative humidity in homes. The average number of mites per square metre is around 200 in a lounge carpet though greater than 10 000 per square metre has been recorded.

Fungal spores and **bacteria** are profuse in heating and ventilating systems. They are thought to create irritation in around one fifth of the population though many of them may not be aware of it.

6.12.3 Radon

Radon is often listed as a pollutant because it is mildly radioactive. It is a product of the decay of uranium in the earth and is found in greater concentrations where specific igneous rocks outcrop, notably Devon, Cornwall and the west coast of Scotland. Although there is evidence of radon in some building materials the geology of the areas where it is a serious problem suggest that the danger is from the local rocks rather than the buildings. It is currently thought to be responsible for 2000 deaths from cancer per year. Without detracting from the seriousness of illnesses caused by radon we should note that by the generally accepted definition of pollution, pollutants are artificial while radon occurs naturally. Radon is therefore better described as a dangerous (carcinogenic) gas than as a pollutant.

6.12.4 Volatile organic compounds (VOCs)

The most common source of VOCs in the home and in the construction industry is usually from paints or similar materials. Modern buildings, especially offices, also have more synthetic material which gives off small amounts of toxic chemicals such as formaldehyde. Concentrations of formaldehyde therefore often depend on the age of the building. Those built before around 1920 contain formaldehyde concentrations of about a third of those built after 1980.

6.13 BUILDING RELATED ILLNESSES (BRI)

It is partly in response to the technological changes in the construction of buildings that we demand more from our homes, offices and factories. Indeed technological change is as much in evidence in the construction industry as it is in other areas of engineering. In commercial buildings especially dependence on artificial ventilation and light and the replacement of natural materials such as timber and plaster with synthetics has resulted in a built environment which has little in common with the natural environment. The difficulties which arise serve to reiterate the impression that isolation from the natural environment has a detrimental effect on our well being.

A building related illness can come about by the occupation of one of two types of buildings, **problem buildings** and **sick buildings**. They are defined by the relationship between the building and illnesses they produce.

6.13.1 Problem buildings

Problem buildings produce symptoms which are directly related to a fault in the construction or design.

- **Example 1** Poorly designed or incorrectly installed ducting can be the cause of minor illnesses such as headaches or nausea. There are many examples of air being extracted from one work area and finding its way into another.
- **Example 2** Following a meeting of the American Legion Convention in Philadelphia in 1976 an outbreak of a pneumonia-like illness occurred among the delegates. It was later found that the bacteria *Legionella pneumophila* was responsible. This particular bacteria lived in the water cooled air conditioning system. The illness was named legionnaire's disease.
- **Example 3** In many old and some relatively new buildings, rising or penetrating dampness and condensation produce an uncomfortable and unhealthy environment. Prolonged exposure to the micro-organisms associated with damp surfaces can lead to bronchial illnesses.

6.13.2 Sick building syndrome (SBS)

Sick buildings produce symptoms which are not related to a particular fault in the building, but which are alleviated by an absence from it. Because of the unknown origin or specific illness, people suffering as result of a sick building are said to suffer from sick building syndrome (SBS). It is worth emphasizing the point that SBS is an illness.

Because of the difficulties involved in relating specific causes to SBS there are many theories on their origins. Some attempts have been made to tie down specific causes, such as microscopic airborne fungi and specific chemicals, while others are prepared to look at more general causes such as recirculating air conditioning systems. Since the specific causes are invariably created by the general causes the latter will be the subject of this section.

The following features put the occupants at risk:

- open plan offices
- permanently sealed draught-proof windows
- water cooled air conditioning
- undiluted recirculation of air
- inappropriate artificial lighting intensity or frequency
- inadequate use of natural lighting.

The items above are features of the building which are dealt with in the design stage. It is not the purpose of this chapter to suggest that architects should not include air conditioning in their specifications. What is the case however is that a combination of those features in the design would almost certainly cost the company occupying the building in terms of lost efficiency and time. Other contributors to SBS which are post-design problems include:

- excessive use of carpets
- excessive shelf space
- redecorating and refurnishing
- accumulation of dust and dust mites
- excessive use of computers
- confrontational management
- low job satisfaction.

6.13.2.1 Causes – sealed building

Section 6.10 showed that a major source of heat loss was air changes. To save fuel, and therefore money, we insulate the building and make it as draught proof as is possible. Modern offices are often designed with sealed windows and an internal environment which is artificially controlled.

The are many natural factors which determine thermal comfort for the individual.

1 People who are physically active or have a higher metabolic rate will usually produce more heat.
2 The amount of heat women produce varies through the menstrual cycle.
3 Eating a large meal increases body heat.
4 Urinating reduces body surface temperature.

The factors listed above tend to suggest that a common environment which cannot be controlled by individuals is not conducive to maximum comfort or efficiency. Future office design may require controls at each work place.

6.13.2.2 Causes – ducted heating and air conditioning

In buildings with no air conditioning natural ventilation through open windows, doors and cracks around frames periodically replaces the internal air.

Air conditioning systems (Figure 6.4) control the air temperature, humidity and cleanness. It is distributed through ducting to the designated space. There are several types. The conventional system consists of a central plant which filters the air and regulates its temperature and humidity. They usually work on a constantly low velocity. The features believed to be associated with SBS are that the plant is centralized and the air is distributed by ducts.

There is a direct relationship between the proportion of fresh air at the mixing chamber and the quality of air in the building. Obviously on a cold day the cost involved in heating outside air makes reducing the intake to zero a financially attractive proposition.

Impurities are removed by the filters and their efficiency depends on original design and proper maintenance. It may be acceptable in some areas for the filter to exclude only insects, but in inner cities the air contains many particles which would not be acceptable inside buildings. In every case the system should include more than one filter. If maintenance is kept to a minimum and the filter fails the material which has been trapped by it since it was installed may find its way back into the room. The most common types are dry screen filters which are simply panels of filter paper or fabric which may be cleaned and returned or disposed of. There is also a roller type which automatically moves onto a clean part of the filter at intervals. There is bound to be a small amount of

the material from the filter screens which makes its way into the room as it rotates. It is therefore important to consider the material from which the screen is manufactured. Those containing, for instance asbestos, should not be used. Wet filters work on much the same principle but are less common.

The most effective are the electrostatic filters. They pass the incoming air through a charged field which gives the particles a electrostatic charge. The collector contains plates with an opposite charge so the particles are attracted to them. Another advantage of electrostatic filters is that they remove bacteria which may, in their absence, be constantly recycled. There is one disadvantage: if the filter is left on with no air running through it the electrostatic field produces ozone from the oxygen in the air. When the air flow is turned on there is a sudden increase in ozone concentration in the air space.

In addition there is an increase in stress and other psychological illnesses which may be either caused or made worse by the open plan design of offices. Many people are uncomfortable with the feeling that they are constantly being observed. SBS may therefore be a response to physical and psychological factors augmented by aggressive or confrontational management techniques, noise, poor decoration, boredom, internal pollutants, lack of control over personal space or lack of comfort.

In the home similar systems are often a major aid to the circulation of minor infectious illnesses such as colds.

There are many assessments of the economic significance of SBS. The cost of absenteeism and reduced efficiency are difficult to calculate. Simple comparisons between one office block and another will produce a very rough guide for the calculation of loss through absenteeism. Current expectation (1995) is that SBS costs the nation around £500 million per annum.

Office workers are most susceptible to SBS. The commonly occurring symptoms are:

● lethargy or fatigue
● loss of concentration
● recurring, ear nose or throat irritations
● infections
● running or blocked nose
● headaches
● irritated eyes
● dry and/or irritated skin.

It is of course possible that all of the above may be the result of some other health factor. While SBS is

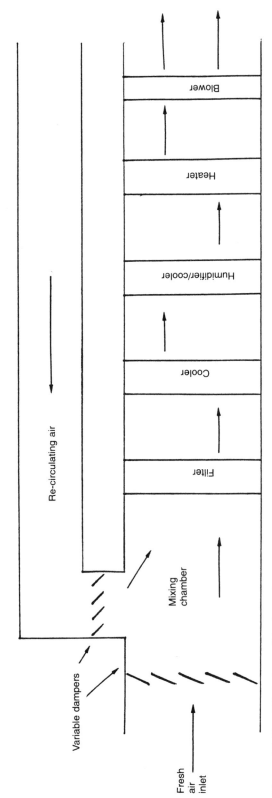

Figure 6.4 Simplified air conditioning unit

neither fatal nor debilitating there is considerable weight of argument suggesting that the buildings we live and work in, and the way we live and work, are making many people ill and costing industries and government large amounts of money. The economic costs are essentially based on productivity, but they include:

- inefficiency caused by minor illnesses
- inefficiency caused by lethargy or drowsiness
- reduction in hours worked per day
- increased absence due to illness
- reduction in time staff spend with a company (hence higher training costs).

The last two items, absence due to illness and staff turnover may have a significant effect on profitability, especially when profit margins are tight.

As well as the public economic cost there is also a personal cost to the people who suffer from SBS. People are not only ill at work. They may convalesce at home for two or three days. If they are constantly ill at home there is no apparent cost to the company but they are likely to become disenchanted, possibly to the extent of changing their job. Government surveys show that productivity is (negatively) affected by SBS and that the best buildings had no air conditioning.

6.14 CONCLUSION

Architects have been charged with the design of buildings which are energy efficient and provide a positive working environment. In some, and by no means all cases, the result has reduced the profitability of the company by creating an unhealthy workforce. In the home a similar scenario has occurred with the removal of the old fashioned firepace and chimney which acted as an excellent ventilator but used a fuel which produces a variety of pollutants. The subsequent introduction of efficient draught proofing and sealed unit double glazing has caused many problems.

It is a vicious circle on a grand scale. To save money and cut pollutants we seal the buildings. This distances us from the natural environment and causes a wide range minor and major illnesses. In Chapter 1 we saw that the original function of the building envelope was to isolate the inhabitants from the natural environment. As construction technology becomes more sophisticated we have become more successful. Unfortunately, we are swapping one set of problems for another and the difficulties are becoming more intricate. Clearly, it is no longer acceptable to consider insulation, ventilation, energy use and pollution as independent topics.

Index